Introduction to Molecular Biology

11th Hour

Introduction to Molecular Biology

Deanna Raineri, PhD
Department of Microbiology
University of Illinois
Urbana, Illinois

Bradley Mehrtens
illustrator and contributor

b
Blackwell
Science

Editorial Offices:
Commerce Place, 350 Main Street, Malden, Massachusetts 02148, USA
Osney Mead, Oxford OX2 0EL, England
25 John Street, London WC1N 2BL, England
23 Ainslie Place, Edinburgh EH3 6AJ, Scotland
54 University Street, Carlton, Victoria 3053, Australia
Other Editorial Offices:
Blackwell Wissenschafts-Verlag GmbH, Kurfürstendamm 57, 10707 Berlin, Germany
Blackwell Science KK, MG Kodenmacho Building, 7-10 Kodenmacho Nihombashi, Chuo-ku, Tokyo 104, Japan

Distributors:
USA
Blackwell Science, Inc.
Commerce Place
350 Main Street
Malden, Massachusetts 02148
(Telephone orders: 800-215-1000 or 781-388-8250; fax orders: 781-388-8270)

Canada
Login Brothers Book Company
324 Saulteaux Crescent
Winnipeg, Manitoba, R3J 3T2
(Telephone orders: 204-837-2987)

Australia
Blackwell Science Pty, Ltd.
54 University Street
Carlton, Victoria 3053
(Telephone orders: 03-9347-0300; fax orders: 03-9349-3016)

Outside North America and Australia
Blackwell Science, Ltd.
c/o Marston Book Services, Ltd.
P.O. Box 269
Abingdon
Oxon OX14 4YN
England
(Telephone orders: 44-01235-465500; fax orders: 44-01235-465555)

Acquisitions: Nancy Whilton
Development: Jill Connor
Production: Louis C. Bruno, Jr.
Manufacturing: Lisa Flanagan
Director of Marketing: Lisa Larsen
Marketing Manager: Carla Daves
Interior design by ColourMark
Cover design by Madison Design
Typeset by Best-set Typesetter Ltd., Hong Kong

Library of Congress Cataloging-in-Publication Data

Raineri, Deanna.
Introduction to molecular biology / by Deanna Raineri.
p.; cm.—(The 11th hour)
ISBN 0-632-04379-2 (alk. paper)
1. Molecular biology—Problems, exercises, etc. 2. Molecular biology—Outlines, syllabi, etc. I. Title. II. 11th hour (Malden, Mass.)
[DNLM: 1. Molecular Biology—Problems and Exercises. QH 506 R155i 2000]
QH506 .R35 2000
572.8—dc21

00-037954

CONTENTS

11TH HOUR GUIDE TO SUCCESS

The 11th Hour Series is designed to be used when the textbook doesn't make sense, the course content is tough, or when you just want a better grade in the course. It can be used from the beginning to the end of the course for best results or when cramming for exams. Both professors teaching the course and students who have taken it have reviewed this material to make sure it does what *you* need it to do. The material flows so that the process keeps your mind actively learning. The idea is to cut through the fluff, get to what you need to know, and then help you understand it.

Essential Background. We tell you what information you already need to know to comprehend the topic. You can then review or apply the appropriate concepts to conquer the new material.

Key Points. We highlight the key points of each topic, phrasing them as questions to engage active learning. A brief explanation of the topic follows the points.

Topic Tests. We immediately follow each topic with a brief test so that the topic is reinforced. This helps you prepare for the real thing.

Answers. Answers come right after the tests; but, we take it a step farther (that reinforcement thing again), we explain the answers.

Clinical Correlation or Application. It helps immeasurably to understand academic topics when they are presented in a clinical situation or an everyday, real-world example. We provide one in every chapter.

Demonstration Problem. Some science topics involve a lot of problem solving. Where it's helpful, we demonstrate a typical problem with step-by-step explanation.

Chapter Test. For more reinforcement, there is a test at the end of every chapter that covers all of the topics. The questions are essay, multiple choice, short answer, and true/false to give you plenty of practice and a chance to reinforce the material the way you find easiest. Answers are provided after the test.

Check Your Performance. After the chapter test we provide a performance check to help you spot your weak areas. You will then know if there is something you should look at once more.

Sample Midterms and Final Exams. Practice makes perfect so we give you plenty of opportunity to practice acing those tests.

The Web. Whenever you see this symbol the author has put something on the Web page that relates to that content. It could be a caution or a hint, an illustration or simply more explanation. You can access the appropriate page through *http://www.blackwellscience.com*. Then click on the title of this book.

The whole flow of this review guide is designed to keep you actively engaged in understanding the material. You'll get what you need fast, and you will reinforce it painlessly. Unfortunately, we can't take the exams for you!

PREFACE

The past 20 years have witnessed an explosion of knowledge in biology, especially at the molecular level. The application of new experimental techniques for the isolation and manipulation of specific fragments of DNA has greatly increased our understanding of the processes responsible for the replication and expression of genetic information. Furthermore, the ability to isolate a specific segment of DNA, modify it in a test tube, and transfer it back into the same or a different cell to create a genetically modified organism represents perhaps the most significant advance in the history of biology, with profound consequences for the future of our, and all, species.

This book covers the molecular processes responsible for the very basis of life itself—DNA replication, transcription, and translation. The related topics of mutagenesis and DNA repair and the control of gene expression in bacteria and eukaryotes are also included. The final chapter introduces the techniques used to manipulate and analyze DNA sequences.

One distinction needs to be made at the outset. Throughout this text, various molecular processes are described as occurring in "bacteria." The "bacteria" to which I refer are the common bacteria (formerly called eubacteria). The archaea or archaebacteria, including the thermophilic (heat-loving), extreme halophilic (salt-loving), and methanogenic (methane-producing) bacteria, are not represented simply because so much still remains to be understood about the basic molecular biology of these ancient organisms.

I have made no attempt to cover every topic that might be included in an introductory molecular biology course. My intention has been to dwell on the more difficult concepts—those that, from my own teaching experience, students tend to find the most difficult to grasp. Comprehensive figures have been included with each chapter for those who are particularly visually oriented, and the Topic and Chapter Tests and Midterm and Final exams provide ample opportunities for testing one's mastery of the material. For a concise review of the material included in an introductory course in general biology, I direct the reader to *Introduction to Biology*, by David Wilson, another book in the 11th Hour Series.

I did not write *Introduction to Molecular Biology* to serve as a stand-alone text. Rather, the student should use it as a supplement to class lecture notes and the more comprehensive textbooks required in introductory biology courses. Students enrolled in majors' or nonmajors' introductory biology courses, those taking advanced placement biology, and preprofessionals studying for the Medical College Admissions Test (MCAT) will benefit from the narratives and exercises provided in this book.

I thank first and foremost Bradley Mehrtens who contributed significantly to this book. I also thank Nancy Whilton, Jill Connor, and Lou Bruno at Blackwell Science for their help, suggestions, and considerable patience. It has been a pleasure working with them. I also thank the following professors and students who reviewed the manuscript and provided helpful comments for improving the legibility and coverage of this study guide: Lizabeth A. Allison, College of William and Mary; Venkat Sharma, University of West Florida; David A. Mullin, Tulane University; Richard Imberski, University of Maryland; Laurie Achenbach, Southern Illinois University; Melody R. Davis, University of Houston; Jessica Koenigsknecht, Baldwin Wallace College; Jill Hung, Baldwin Wallace College; Alejandra

Gutierrez, University of Houston. Finally, I thank my husband Tom Jacobs for his support and encouragement.

This book is dedicated to my parents, Jacqueline and Giuseppe Raineri.

Deanna M. Raineri, Ph.D.
Professor of Biology
University of Illinois

CHAPTER 1

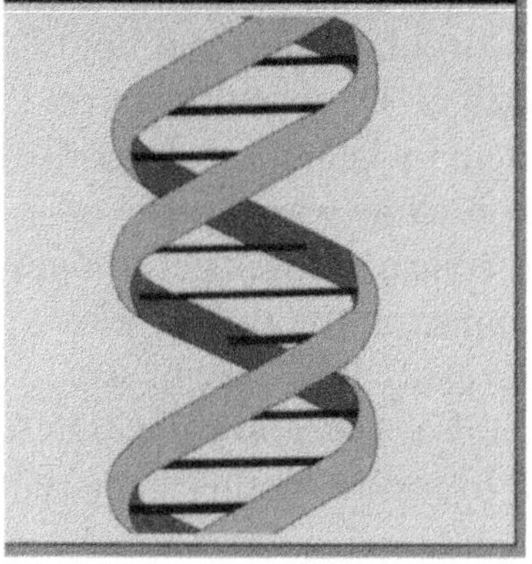

DNA: The Genetic Material

By the early 1940s, hereditary information was known to reside on chromosomes, which consist of proteins and DNA. Because proteins are so heterogeneous, with 20+ different kinds of amino acids commonly found in proteins compared with just 4 different units that comprise DNA, most scientists believed that proteins were the genetic material. That DNA is the genetic material was established through work with bacteria and bacteriophages (denoted as phages), viruses that infect and reproduce in bacteria.

ESSENTIAL BACKGROUND

- **Basic chemistry of biological molecules**
- **Properties of hydrogen bonding**
- **Packaging of DNA into chromosomes**

TOPIC 1: SEARCH FOR THE GENETIC MATERIAL

KEY POINTS

✓ *What properties define genetic material?*

✓ *How did scientists prove that DNA is the genetic material?*

VIDENCE THAT DNA CAN ALTER PROPERTIES OF CELLS

Frederick Griffith worked with two strains of *Streptococcus pneumoniae*—a pneumonia-causing "smooth" or "S" strain (so-called because it synthesizes a thick outer polysaccharide capsule that gives the colonies a smooth appearance) and a mutant strain that does not cause pneumonia, called the "rough" or "R" strain (so-called because it lacks the outer polysaccharide coat giving the colonies a rough appearance). When Griffith injected mice with live cells of the pneumonia-causing S strain, the mice died of pneumonia as expected. However, the mice also died when Griffith injected mice with a mixture of heat-killed S cells and live R cells, and, even more surprising, **live S cells** could be isolated from the dead mice. Because neither component of this mixture alone (either the heat-killed S cells or the live but harmless R cells) caused pneumonia when injected into mice, Griffith concluded that genetic material released from the dead S cells was incorporated into the live R cells, "transforming" them into S-type pneumonia-causing bacteria. Griffith called this process **transformation**, but he himself did not identify the nature of the transforming agent. This task was accomplished some 16 years later in 1944 by Oswald T. Avery, MacLyn McCarty, and Colin MacLeod. Avery and his colleagues repeated Griffith's experiments, but they subjected the extract from heat-killed S cells to various treatments to selectively destroy the different types of molecules (proteins, nucleic acids, carbohydrates, and lipids) present in the extract. In so doing, they found that the selective removal of DNA with deoxyribonuclease was the only treatment that resulted in the transforming activity being lost. In addition, they isolated pure DNA from heat-killed S cells and showed that it alone was capable of transforming R cells into S cells. Although the work of Avery and his colleagues was paramount in establishing that DNA is the genetic material in cells, it had little impact at the time it was published. Some criticized the experimental data arguing that the "purified" DNA might have been contaminated with protein, whereas others pointed out that even if *S. pneumoniae* uses DNA as its genetic material, this did not prove that animals and plants, not to mention other bacteria, also use DNA.

FURTHER EVIDENCE THAT DNA AND NOT PROTEIN IS THE GENETIC MATERIAL: THE ABILITY OF PHAGE DNA TO REPROGRAM HOST CELLS

In 1952, Alfred D. Hershey and Martha Chase provided further evidence to support Avery et al.'s claims by demonstrating that DNA is the genetic material of a phage known as T2. T2 has a very simple structure, consisting of little more than DNA surrounded by a protein coat. T2 reproduces by inserting its genetic material into its host cell, the bacterium *Escherichia coli* and completely commandeering the host cell's resources.

Hershey and Chase devised a way to differentiate between the DNA and the protein of the phage T2 through the use of radioactive isotopes. Their approach is based on the observation that all proteins contain some sulfur (in the amino acids cysteine and methionine), whereas sulfur is not present in DNA. On the other hand, DNA is rich in phosphorus, an element that is not present in most proteins. Thus, by growing one batch of T2 in the presence of a radioactive isotope of sulfur (^{35}S) and another batch of T2 in the presence of a radioactive isotope of phosphorus (^{32}P), two different populations of T2 cells were obtained—one carrying ^{35}S-labeled proteins and the other carrying ^{32}P-labeled DNA. Different *E. coli* cultures were then infected with the labeled T2 cells, the cultures blended in a kitchen blender to dislodge any parts of the phage remaining on the surface of the *E. coli* cells, and the mixtures centrifuged to separate the bacterial cells from the lighter phage particles. In the culture with the radioactively labeled phage proteins, most of the radioactivity remained in the supernatant, suggesting that the phage proteins did not enter the *E. coli* cells. In the culture with the radioactively labeled phage DNA, most of the radioactivity was detected in the bacterial pellet, suggesting that it was the phage DNA that entered the *E. coli* cells and was therefore the genetic material.

CIRCUMSTANTIAL EVIDENCE THAT DNA IS THE GENETIC MATERIAL

Further support for the role of DNA as the genetic material of cells came from the observation that the amount of DNA doubles immediately before (each) cell division and that diploid cells (in eukaryotes) contain twice as much DNA as haploid gametes, properties that are consistent with genetic material.

Topic Test 1: Search for the Genetic Material

True/False

1. The observation that haploid gametes contain half the amount of DNA as diploid cells provides additional evidence that DNA is the genetic material.

Multiple Choice

2. From their experimental data, Hershey and Chase concluded that
   a. The ability of DNA to move between cells is inhibited by certain enzymes.
   b. ^{35}S selectively labels proteins and ^{32}P selectively labels DNA.
   c. Most of the ^{35}S, and not the ^{32}P, enters the infected cells.
   d. The phage DNA, not the phage protein, enters the host cell.
   e. A newly replicated molecule of DNA contains one strand from the parent molecule and one newly synthesized strand.

3. Which one of the following graphs most accurately reflects the conclusions drawn from the Hershey–Chase experiment?

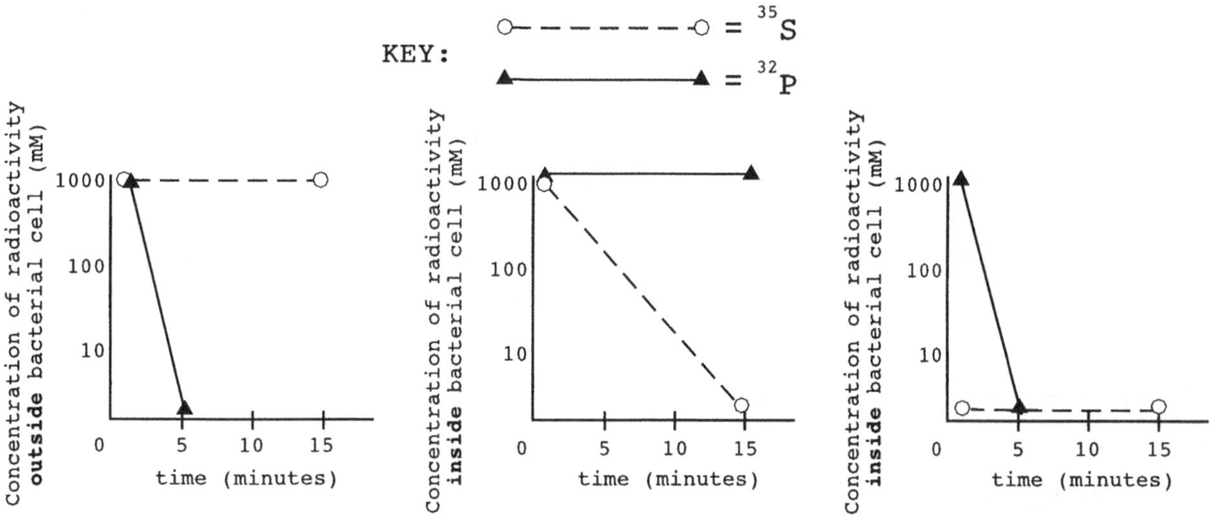

A B C

Short Answer

4. When Griffith injected mice with a mixture of heat-killed S cells and live R cells of *S. pneumoniae* bacteria, why did the mice die?

5. Until DNA was shown to be the genetic material, why was protein thought to be the most likely candidate?

Topic Test 1: Answers

1. **True.** The observation that haploid cells have half the DNA of diploid cells may have been circumstantial. However, when combined with the experimental evidence obtained by Griffith and Hershey and Chase, the observation that haploid cells have half the DNA of diploid cells helped to confirm that DNA is indeed the genetic material.
2. **d. The phage DNA, not the phage protein, enters the host cell.** Hershey and Chase were trying to identify the molecule that comprises the genetic material. By using bacteriophage, viruses that are known to inject their genetic material into bacteria, they selectively labeled the phage protein with ^{35}S and the phage DNA with ^{32}P. After allowing the phage to deliver their genetic material, the bacteria were separated from the phage. They found that most of the ^{32}P, and thus the phage DNA, remained with the bacterial fraction. This proved that it was the phage DNA that was transferred into the bacterial cells and therefore that DNA is the genetic material.
3. **Graph A.** Hershey and Chase demonstrated that DNA is the genetic material by showing that it is the DNA that is injected from a bacteriophage into the bacterial cell. They did this by labeling the DNA of the phage with a radioactive isotope of phosphorus (^{32}P) and the protein coat of the phage with a radioactive isotope of sulfur (^{35}S) (phosphorus is not found in most proteins and there is no sulfur in DNA). Most of the ^{32}P, and therefore the phage DNA, passed into the bacterial cell, whereas the ^{35}S, and therefore the phage protein, remained outside the cell. Only Graph A meets these criteria.
4. The mice died, even though the only virulent bacteria injected into them were heat killed, because the genetic material from the virulent S bacteria was still intact. The genetic material from the dead S cells was taken up by the live nonvirulent R cells. The nonvirulent R cells then expressed the genetic material from the S cells, including the gene for the outer polysaccharide coat that allows the bacteria to evade the immune system of the mice, thereby "transforming" the R cells into virulent S-type bacteria.
5. Protein was long believed to be the genetic material because it was known that 20 different amino acids are found in protein but only 4 different nucleotides are found in DNA. Proteins therefore have a much greater potential for variation in composition than DNA and hence seemed the better candidate to code for the incredible genetic diversity found in life.

TOPIC 2: DNA STRUCTURE

KEY POINTS

✓ *What is DNA composed of?*

✓ *What were the most critical observations that led to the discovery of the three-dimensional structure of DNA?*

✓ *How was the three-dimensional structure of DNA determined?*

COMPOSITION OF DNA

DNA is a nucleic acid. The nucleic acids (DNA and RNA) are polymers consisting of fewer than a hundred to millions or even billions of monomeric units called nucleotides. Each nucleotide

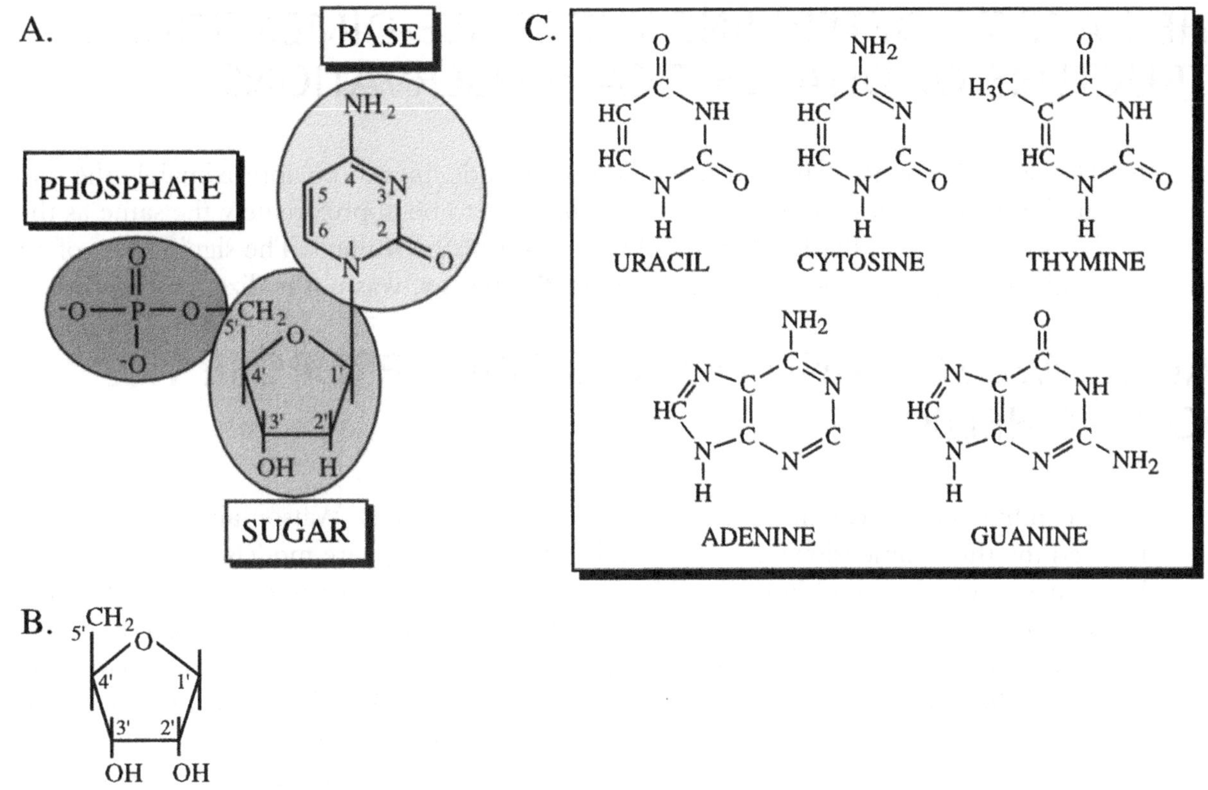

Figure 1.1 Structure of nucleotides. (A) Components of a DNA nucleotide, including a five-carbon sugar (deoxyribose in DNA), a phosphate group, and a nucleotide base that varies according to nucleotide type. (B) Ribose, the five-carbon sugar that is found in RNA in place of the deoxyribose shown in A. (C) Nucleotide bases commonly found in DNA and RNA. Uracil, cytosine, and thymine are referred to as pyrimidines; adenine and guanine are purines.

consists of a five-carbon sugar (a pentose) covalently bonded to a phosphate group and a nitrogen-containing **purine** or **pyrimidine** base (**Figure 1.1**).

Two types of pentose sugar are found in nucleic acids. **Ribose** is the pentose found in all RNAs, whereas **deoxyribose**, in which the hydroxyl group at the 2′ position in ribose is replaced by a hydrogen, is found in all DNAs. DNA and RNA also differ in their nucleotide base composition. There are two families of nucleotide bases: the **pyrimidines**, including **cytosine (C)**, **thymine (T)**, and **uracil (U)**, are characterized by a single six-membered ring of carbon and nitrogen atoms, whereas the **purines**, including **adenine (A)** and **guanine (G)**, consist of two-ring structures, a six-membered ring attached to a five-membered ring of carbon and nitrogen atoms. The purines, adenine and guanine, are found in both nucleic acids, but the pyrimidine composition of DNA differs from that of RNA. Thus, cytosine is found in both DNA and RNA, whereas thymine is only found in DNA and uracil is only found in RNA.

DNA and RNA are synthesized from triphosphate derivatives of the above-mentioned nucleotides (so-called deoxyribonucleoside triphosphates, abbreviated dNTP, where "N" is A, G, C or T). Energy released by removal of the two terminal phosphate groups (pyrophosphate) from each nucleotide drives the formation of a phosphodiester linkage that joins the 5′-phosphate of one nucleotide to the 3′-OH group on the sugar of the next nucleotide.

THE RACE TO DETERMINE THE THREE-DIMENSIONAL STRUCTURE OF DNA: CRITICAL OBSERVATIONS

In 1947 Erwin Chargaff determined the number of adenines to be approximately the same as the number of thymines and the number of guanines to be approximately the same as the number of cytosines in the DNA from all the organisms he studied. The significance of these observations, which became known as **Chargaff's Rules**, was not realized at the time.

JAMES WATSON AND FRANCIS CRICK PROPOSE THE DOUBLE HELIX

Using combined chemical and physical data for DNA, James D. Watson and Francis Crick deduced the three-dimensional structure of DNA by building wire models to conform with x-ray data generated by Rosalind Franklin. These x-ray data were generated by bombarding DNA fibers with an x-ray beam. X-rays scattered by the fibers exposed a photographic film and, in so doing, produced a pattern that gave information about the three-dimensional structure of DNA. The x-ray photograph generated by Franklin suggested that DNA is a helical molecule with the purine and pyrimidine bases of the nucleotides stacked one on top of the other like a pile of plates (giving rise to so-called **base stacking** interactions). Putting the x-ray data generated by Franklin together with everything they knew about the chemical and physical properties of DNA, Watson and Crick suggested that DNA actually exists as a **double helix**, with two separate chains of DNA wound around each other (**Figure 1.2**). The uniform 2 nm width of the helix suggested that a purine base on one DNA strand must be opposite a pyrimidine base on the other DNA strand, and vice versa. Watson and Crick finally arrived at a model that placed the sugar-phosphate backbones of the DNA chains on the outside of the helix and the nucleotide bases on the inside. They further concluded that the two strands of the double helix are oriented in opposite directions, that is, they are **antiparallel**. In this orientation, the molecular arrangements of the side groups of the nucleotide bases on the opposing DNA chains allow two hydrogen bonds (H bonds) to form between adenine and thymine and three H bonds between guanine and cytosine. This finding unexpectedly explained Chargaff's observations that the amount of adenine equals the amount of thymine and the amount of guanine equals the amount of cytosine in any DNA because adenine is always "base paired" to thymine and guanine is always "base paired" to cytosine. This specific pairing of the nucleotide bases is known as **complementary base pairing** (Figure 1.2). The complementary base pairs are rotated 36 degrees with respect to each adjacent base pair, giving 10 base pairs for each 360-degree turn of the helix.

THE DOUBLE HELIX CAN BE DENATURED AND RENATURED

Many of the molecular techniques used to characterize DNA require that the double helix is separated into its individual strands. Because the purine and pyrimidine bases are adjacent in a DNA chain, **base stacking** gives some rigidity to the sugar-phosphate backbones so that instead of forming a random coil structure, the DNA chains remain extended. Base stacking is therefore of major importance in stabilizing the helical structure of DNA and the double-stranded structure of the helix is maintained by hydrogen-bonding between the bases of the

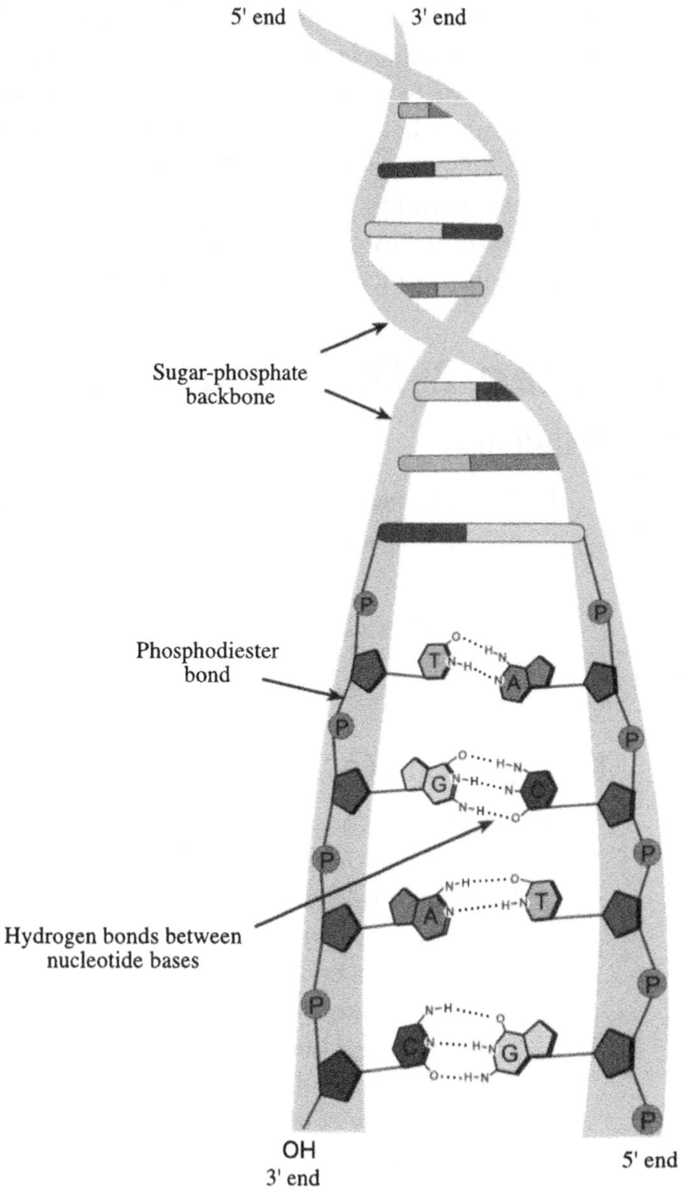

Figure 1.2 DNA is a double helix held together by complementary base pairs. The sugar and phosphate residues of the DNA nucleotides make up the backbone of the double helix. The nucleotide bases are located on the inside of the double helix where they help to stabilize the helix by forming complementary base pairs held together by hydrogen bonds.

complementary base pairs. Any treatment that will disrupt the base stacking interactions and break the hydrogen bonds between the complementary base pairs, such as exposure to high temperatures or treatment with acids or bases, will cause the two DNA strands to separate, while leaving the covalent bonds, including the phosphodiester bonds, between the individual nucleotides intact. This process is known as **DNA denaturation**.

The higher the percentage of GC base pairs, the greater the number of hydrogen bonds that need to be broken and hence the higher the temperature needed to denature the double helix. Similarly, higher temperatures are needed to denature a double helix in the presence of salt. Positively charged ions, such as Na^+, K^+, and Li^+, neutralize the instability caused by the negative charges carried by the phosphate groups in the DNA backbone, thereby eliminating electrostatic repulsion between the phosphates. Thus, positively charged ions make the double helix more stable (and hence more difficult to denature).

Denatured DNA can reform double-stranded DNA, a process called **renaturation**. Renaturation is a very valuable tool in molecular biology. For example, it can be used to determine how closely related two different organisms are as homologous sequences from two different DNAs can renature when they are denatured and mixed together. The extent of "cross-hybridization" between different denatured DNAs gives information about the degree of relatedness between different organisms. Renaturation can also be used to determine how often a particular sequence occurs in the DNA of an organism because repetitive DNA has a greater chance of finding a homologous partner and hence of renaturing than single-copy DNA.

Topic Test 2: DNA Structure

True/False

1. The statement that the DNA double helix is "antiparallel" means that all the purine bases are on one side and all of the pyrimidine bases are on the other side.
2. In addition to the deoxyribose sugar and phosphate groups that comprise the backbone, RNA strands contain the nucleotide bases adenine, guanine, cytosine, and uracil.

Multiple Choice

3. Adenines account for 15% of the nucleotide bases in the DNA from organism A and guanines account for 35% of the nucleotide bases in the DNA from organism B. Use Chargaff's rules to determine which one of the following statements is correct.
   a. Guanines account for 70% of the nucleotide bases in organism A's DNA and adenines account for 65% of the nucleotide bases in organism B's DNA.
   b. Guanines account for 35% of the nucleotide bases in organism A's DNA and adenines account for 30% of the nucleotide bases in organism B's DNA.
   c. Guanines account for 35% of the nucleotide bases in organism A's DNA and adenines account for 15% of the nucleotide bases in organism B's DNA.
   d. It would probably take more energy to denature DNA from organism A than to denature DNA from organism B.
   e. It would probably take less energy to denature DNA from organism A than to denature DNA from organism B.
4. Given the three-dimensional structure of DNA as described by Watson and Crick, which of the following is true of the DNA double helix (more than one of the following statements may be correct)?
   a. The total number of purine bases equals the total number of pyrimidine bases.
   b. The 5′ end of one of the DNA strands is aligned with the 5′ end of the other DNA strand.
   c. Adenines always occur opposite thymines and guanines always occur opposite cytosines.
   d. The sequence of bases on one DNA strand is identical to the sequence of bases on the other DNA strand.
   e. The sequence of bases on one DNA strand is complementary to the sequence of bases on the other DNA strand.

Short Answer

5. Of the two different DNA molecules below, which one would you expect to denature at a lower temperature and why?
   a. GCATTGCCAATGC
   b. ATTAGCCTATCGG

6. If a 100 base pair DNA double helix contains 45 cytosines, how many of each of the other nucleotide bases does this DNA contain?

Topic Test 2: Answers

1. **False.** The statement that the DNA double helix is "antiparallel" means that the two DNA strands are oriented in opposite directions. In other words, the nucleotides of one of the DNA strands are upside-down relative to the nucleotides of the other DNA strand.

2. **False.** Adenine, guanine, cytosine, and uracil are the nucleotide bases found in RNA, but RNA contains **ribose** as its pentose sugar, not deoxyribose.

3. **c. Guanines account for 35% of the nucleotide bases in organism A's DNA and adenines account for 15% of the nucleotide bases in organism B's DNA.** Adenine is always base paired with thymine and guanine is always base paired with cytosine. Therefore, the number of adenines in any organism's DNA must equal the number of thymines and the number of guanines must equal the number of cytosines. If adenines account for 15% of organism A's DNA, then thymines must also account for 15%. This leaves the remaining 70% to be split equally between guanine and cytosine (35% each). Similarly, if guanines account for 35% of organism B's DNA, then cytosines must also account for 35%. This leaves 30% to be split equally between adenine and thymine (15% each). Only c matches these criteria. Neither d nor e is correct because the organisms have identical guanine–cytosine content and would therefore be expected to denature at approximately the same temperature.

4. **a, c, and e are correct.** Because every purine (adenine or guanine) is base paired with a pyrimidine (thymine or cytosine) in a double helix, the number of purines must equal the number of pyrimidines. b is not correct because the strands are *antiparallel*—the 5′ end of one strand matches up with the 3′ end of the other strand and vice versa. c correctly describes complementary base pairing, and because base pairing is **complementary**, it follows that e is also correct, whereas d is not.

5. **Molecule b would be expected to denature at a lower temperature.** GC base pairs are held together with three hydrogen bonds, whereas AT base pairs are held together with only two. It therefore takes more energy (i.e., heat) to break a GC bond than an AT bond. Molecule B has only six GC bonds, and molecule A has seven, so a lower temperature probably would be required to denature molecule B.

6. **A 100 base pair double helix has a total of 200 nucleotide bases.** If 45 are cytosines, by Chargaff's rules, 45 are guanines. That accounts for 90 out of 200, leaving 110 to be divided equally between adenine and thymine, for a total of 55 each.

TOPIC 3: DNA REPLICATION

KEY POINTS

- ✓ *Why does complementary base pairing immediately suggest a mechanism for DNA replication?*
- ✓ *How did Meselson and Stahl prove that DNA replicates semiconservatively?*
- ✓ *Why does the antiparallel nature of the double helix pose a problem during DNA replication?*

An adenine in one of the DNA strands of the double helix is always opposite a thymine in the other DNA strand and vice versa. Similarly, guanines are always opposite cytosines and cytosines opposite guanines. These complementary base-pairing rules suggest a mechanism for DNA replication because, if separated, each DNA strand could serve as a template for the rebuilding of a new complementary strand. Three models for DNA replication were proposed.

The **semiconservative** model predicts that the double helix unwinds to generate two new daughter molecules, each containing one of the original parental strands and one newly synthesized strand. In the **conservative** model, the parental helix remains intact and passes into one of the daughter cells; the other daughter cell inherits a double helix consisting of two newly synthesized DNA strands. In the **dispersive** model, all four strands of the two daughter helices are a mixture of parental and new DNA.

To determine which of the three models is correct, Matthew Meselson and Franklin Stahl grew *E. coli* in a medium containing a heavy isotope of nitrogen (^{15}N). After several generations, the heavy isotope was incorporated into the nucleotide bases and the bacterial DNA molecules became "heavy." Cells with this heavy DNA were then transferred into medium containing the normal "light" isotope of nitrogen (^{14}N) and allowed to replicate. After one cell division, the DNA was extracted and found to be of *intermediate* density ($^{15}N/^{14}N$), suggesting that both new daughter molecules were hybrid molecules, each consisting of a mixture of parental and newly synthesized DNA. After two cell divisions in the light medium, two distinct DNA populations were observed; half of the cells contained only "light" DNA ($^{14}N/^{14}N$) and the other half contained DNA of intermediate density ($^{15}N/^{14}N$). These results were consistent with the **semiconservative** model (see the Demonstration Problem at the end of this chapter for an explanation of this experimental data).

DNA REPLICATION REQUIRES A TEAM OF ENZYMES

DNA Replication Starts at Specific Sites Called Origins of Replication

Proteins that initiate replication bind to a specific sequence of nucleotides (the origin of replication or "*ori*") and separate the two DNA strands to form a "replication bubble." Enzymes called **helicases** then unwind the helix in both directions away from the origin of replication, resulting in two Y-shaped **replication forks**. **Single-strand DNA binding proteins** then coat the two DNA strands to keep them apart.

Because most bacterial chromosomes are small and circular, they can get by with only one origin of replication. In contrast, to be able to replicate the long linear DNA molecules of eukaryotic chromosomes in a reasonable time frame, multiple *ori* are required. Where replication forks from adjacent replication bubbles meet, they fuse to form a completely replicated piece of DNA (**Figure 1.3**).

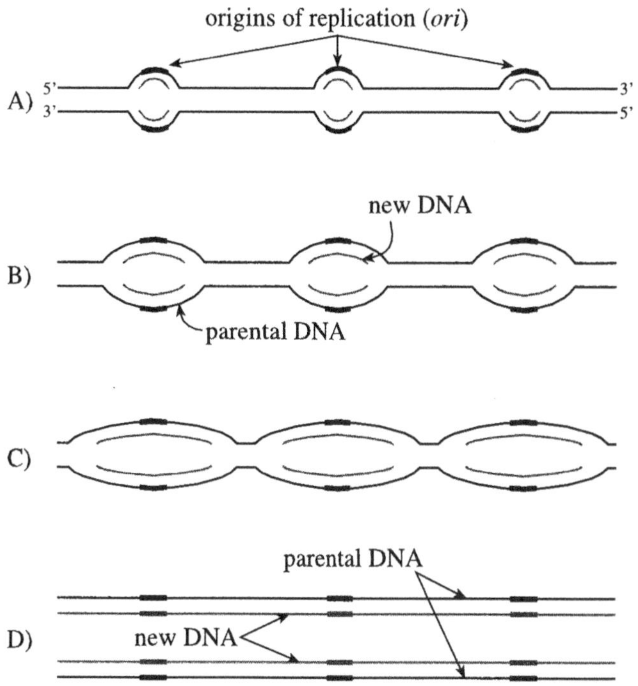

Figure 1.3 Origins of replication in eukaryotes. Multiple origins of replication (*ori*) are needed to replicate eukaryotic chromosomes. The replication bubbles resulting from these *ori* eventually join, giving rise to two new complete DNA strands.

Replication of Circular DNA Introduces Twists

As improvements were made in the techniques used to isolate DNAs, it became clear that many bacterial DNA molecules are circular with the ends of the double helix covalently joined together. The circular nature of bacterial genomes presents a formidable topological problem during DNA replication. Thus, for a replication fork to move along a piece of DNA, the DNA helix would need to unwind ahead of it, causing the DNA to rotate. However, because the bacterial chromosome is circular, there are no free ends to rotate. Therefore, in the absence of some kind of swiveling, the unreplicated portion of the double helix would become overwound and get very tight, eventually restricting any further advancement of the replication fork. Although not circular, eukaryotic nuclear genomes encounter a similar topological problem during replication. This is due to the fact that eukaryotic nuclear genomes initiate replication simultaneously at multiple *ori*, resulting in long DNA chains of many small replicating circles—analogous to replicating bacterial genomes.

Overwinding of the DNA helix during replication is prevented by enzymes called **topoisomerases**. Together these enzymes modify the superhelicity of circular DNA by breaking one or both DNA strands and twisting the strands with respect to one another before linking them together again. This has the effect of unwinding or "relaxing" the DNA.

Elongation: Directionality of DNA Synthesis

Each DNA strand has polarity. The deoxyribose sugar of each DNA nucleotide is attached to a phosphate group at its 5′ carbon atom and to a hydroxyl group at its 3′ carbon atom. When a nucleotide is added to a growing DNA chain, the phosphate group of the new nucleotide bonds to the hydroxyl group attached to the 3′ carbon of the nucleotide at the end of the chain. In other words, nucleotides are always added to the same end of the growing DNA strand—the 3′ end (**Figure 1.4**).

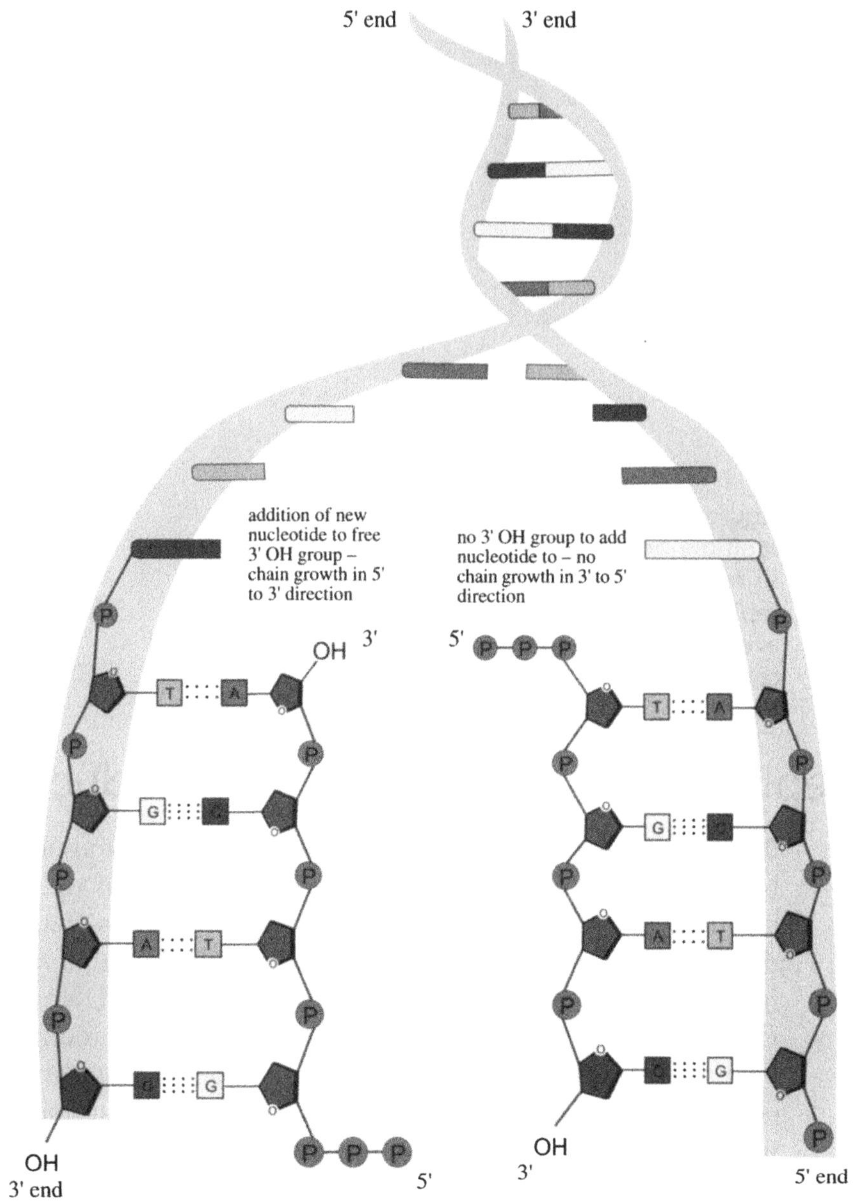

Figure 1.4 Directionality of DNA synthesis. DNA can only be synthesized in a 5′ to 3′ direction because new nucleotides are always added to the 3′-OH group of the last nucleotide in the chain.

PROBLEM OF ANTIPARALLEL STRANDS

In semiconservative replication, the helix separates at an origin of replication site and the parental DNA on *both* sides of the *ori* is replicated. However, if DNA synthesis can only occur in one direction, how is the parental template replicated on both sides of the *ori* because, at first glance, this would seem to indicate that DNA synthesis would occur in a 3′ to 5′ and in a 5′ to 3′ direction as follows:

```
                      ori
5′---------------------ΞΞ---------------------3′ Parental DNA
            3′<------5′ 3′------>5′?              New DNA
```

On one side of each *ori* the parental strand runs in a 3′ to 5′ direction away from the *ori* toward the replication fork. Because the two strands of a DNA helix are antiparallel, to replicate the parental DNA on this side of the *ori*, the new DNA would have to be synthesized in the 5′ to 3′ direction, which is the correct direction for DNA synthesis. This mode of DNA replication is referred to as *continuous* replication and the resulting new DNA strand as the **leading strand** (**Figure 1.5**).

The problem occurs on the other side of each *ori*, where the parental strand runs in a 5′ to 3′ direction away from the *ori* toward the replication fork. To replicate the DNA on this side of the *ori*, the new DNA strand would need to be synthesized in a 3′ to 5′ direction if replication started at the *ori*, but no DNA polymerases have been found that are capable of adding new nucleotides to the 5′ end of a DNA chain. In this case, replication does not start at the *ori*. Instead, in *E. coli*, replication of the parental strand on the 5′ to 3′ side of the *ori* does not begin until 1,000–2,000 nucleotides of the parental strand have been exposed by unwinding (100–200 nucleotides in eukaryotes). Synthesis of the new complementary DNA is then initiated, not at the *ori* but close to the replication fork and chain growth proceeds in the correct 5′ to 3′ direction *away* from the replication fork (i.e., *toward* the *ori*) to form the first **Okazaki fragment**, named after its discoverer, Reiji Okazaki. As the replication fork progresses, exposing another 1,000–2,000 nucleotides of the parental template, a second Okazaki fragment is formed and so on. This mode of replication is referred to as *discontinuous* replication and the resulting fragmented strand as the **lagging** strand (Figure 1.5).

DNA POLYMERASES AND THE NEED FOR RNA PRIMERS

Enzymes called DNA polymerases are responsible for synthesizing new DNA. However, the DNA polymerases by themselves cannot initiate DNA synthesis; they can only add nucleotides to a 3′-OH group at the end of an existing chain. In *E. coli*, this problem is resolved by two different RNA-synthesizing enzymes—**RNA polymerase** (which is responsible for the synthesis of most RNA molecules in *E. coli* as explained in Chapter 2) and **primase**. Functionally, these enzymes are indistinguishable—they pair *RNA nucleotides* to 10–20 DNA bases to form a short chain of RNA. This short RNA chain acts as a primer for *E. coli* **DNA polymerase III** by providing it with a 3′-OH group. The difference between RNA polymerase and primase lies in how many primers they have to make. Thus, RNA polymerase is responsible for initiation of leading strand synthesis. Because the leading strand grows continuously, it only needs to be primed once. Primase, on the other hand, is responsible for the initiation of each of the Okazaki fragments. Primase therefore has to synthesize primers repetitively as the lagging strand is elongated. In eukaryotes, the RNA primers are made by one of the DNA polymerases (DNA polymerase α) and they are extended by DNA polymerase δ.

The short RNA chains used to prime DNA synthesis have to be removed and replaced by DNA. In *E. coli*, this is achieved by **DNA polymerase I**, which attaches to the 3′ end of an Okazaki fragment and synthesizes DNA in the 5′ to 3′ direction. When the DNA polymerase I meets an RNA primer, it removes the RNA nucleotides one at a time and replaces them with DNA. In eukaryotes, two different enzymes are required: a ribonuclease excises the RNA primer and DNA polymerase δ synthesizes DNA in its place. (For a comparison of bacterial and eukaryotic DNA polymerases, see **Table 1.1**.)

When all RNA primers have been replaced by DNA, small nicks remain in the sugar-phosphate backbone of the new DNA strand. These nicks are sealed by **DNA ligase** to form a complete continuous DNA strand.

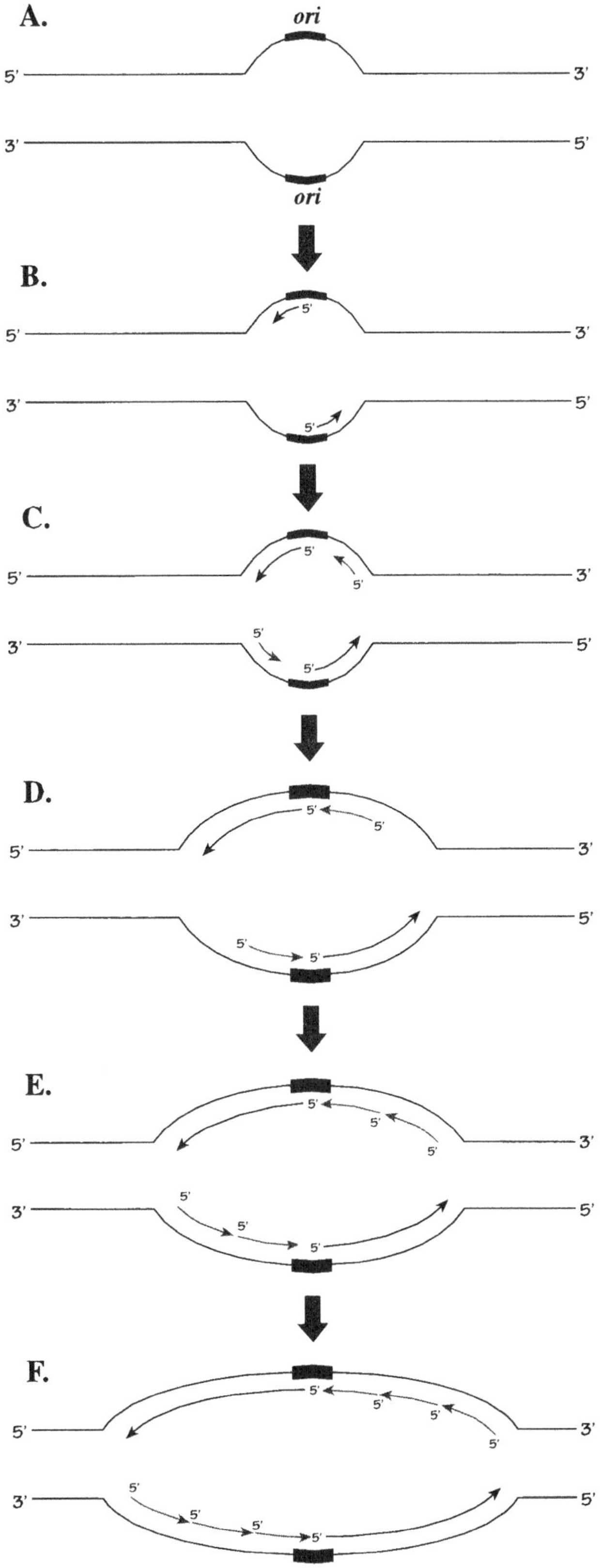

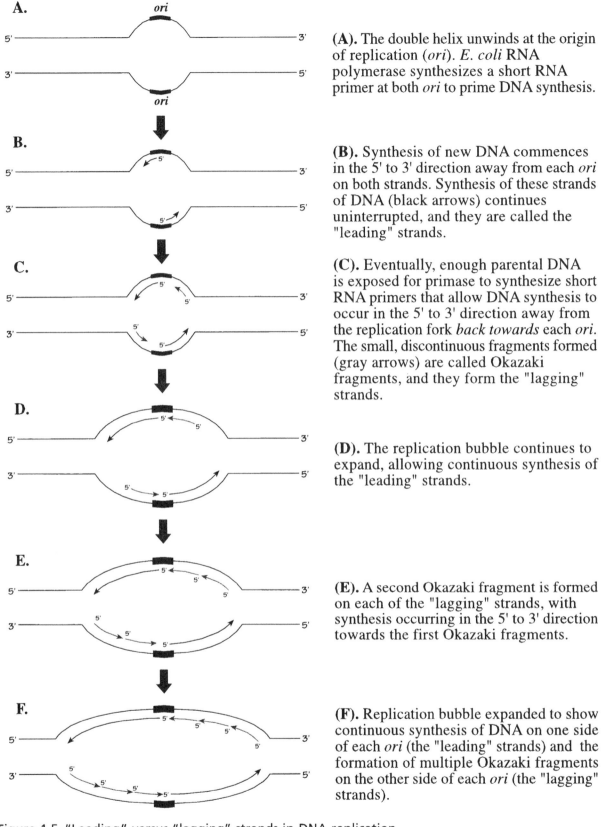

(A). The double helix unwinds at the origin of replication (*ori*). *E. coli* RNA polymerase synthesizes a short RNA primer at both *ori* to prime DNA synthesis.

(B). Synthesis of new DNA commences in the 5' to 3' direction away from each *ori* on both strands. Synthesis of these strands of DNA (black arrows) continues uninterrupted, and they are called the "leading" strands.

(C). Eventually, enough parental DNA is exposed for primase to synthesize short RNA primers that allow DNA synthesis to occur in the 5' to 3' direction away from the replication fork *back towards* each *ori*. The small, discontinuous fragments formed (gray arrows) are called Okazaki fragments, and they form the "lagging" strands.

(D). The replication bubble continues to expand, allowing continuous synthesis of the "leading" strands.

(E). A second Okazaki fragment is formed on each of the "lagging" strands, with synthesis occurring in the 5' to 3' direction towards the first Okazaki fragments.

(F). Replication bubble expanded to show continuous synthesis of DNA on one side of each *ori* (the "leading" strands) and the formation of multiple Okazaki fragments on the other side of each *ori* (the "lagging" strands).

Figure 1.5 "Leading" versus "lagging" strands in DNA replication.

Table 1.1 Bacterial versus Eukaryotic DNA Replication

	BACTERIA (E. coli)	EUKARYOTES
*ori**	**one**	**multiple**
Strand separation/strand coating	DNA helicases/single-strand binding proteins	DNA helicases/single-strand binding proteins
Primer synthesis	RNA polymerase (leading strand) DNA primase (lagging strand)	DNA polymerase α
Strand elongation	DNA polymerase III	DNA polymerase δ
Primer removal	DNA polymerase I	Ribonuclease
Primer replacement	DNA polymerase I	DNA polymerase δ
Sealing gaps	DNA ligase	DNA ligase

* Origins of replication.

DNA POLYMERASES CAN REMOVE MISPAIRED NUCLEOTIDES

The *E. coli* DNA polymerases make about one error for every 10^8 nucleotides replicated. This error rate is so low partly because both DNA polymerases have the ability to correct any mistakes they make. Thus, in addition to catalyzing the 5′ to 3′ synthesis of DNA, *E. coli* DNA polymerases I and III and some eukaryotic DNA polymerases possess a **3′ to 5′ exonuclease** or **proofreading** activity that allows them to go back over what they have just synthesized and correct any errors. Thus, before adding the next nucleotide to a growing DNA chain, each DNA polymerase checks to see whether the last nucleotide added is correctly base paired to the template strand. If the last nucleotide is mispaired, the DNA polymerase breaks the phosphodiester bond it just made (thereby releasing the mispaired nucleotide) and tries again.

Topic Test 3: DNA Replication

True/False

1. A DNA polymerase recognizes specific sequences in the parental DNA and unwinds the double helix at these sites so that DNA replication can occur.
2. The 3′ to 5′ exonuclease activity of *E. coli* DNA polymerases I and III allows the polymerases to remove misincorporated nucleotides, thereby reducing the number of errors made during DNA replication.
3. Most eukaryotic DNAs contain multiple origins of replication (*ori*), whereas most bacterial DNAs contain only one *ori*.
4. If the two DNA strands of a double helix are read in the same 5′ to 3′ direction, the sequence of nucleotides would be identical.

Multiple Choice

5. In *E. coli*, a DNA polymerase lacking a 5′ to 3′ exonuclease activity would not be able to
   a. synthesize the leading strand.

b. repair the nicks left between Okazaki fragments.
c. remove misincorporated nucleotides.
d. remove RNA primers.
e. None of the above.

6. If the structure of DNA was compared with a spiral staircase, which of the following would be analogous to the steps of the staircase?
   a. Linked deoxyribose and phosphate groups.
   b. Purine–pyrimidine base pairs.
   c. Linked ribose and phosphate groups.
   d. Adenine–guanine base pairs.
   e. Guanine–thymine base pairs.

7. Which of the following events occurs during DNA replication in *E. coli*?
   a. Removal of Okazaki fragments by DNA polymerase III.
   b. Discontinuous synthesis of DNA on the leading strand.
   c. Synthesis of short RNA primers.
   d. Degradation of the old (parental) double helix.
   e. Unwinding of the double helix by DNA polymerase I.

Short Answer

8. What is meant by "semiconservative" DNA replication?

9. A mutant bacterial cell contains a defective chromosome. The chromosome consists of a normal DNA strand base paired with numerous small DNA fragments. Which enzyme is **most likely** to be missing or defective in this mutant cell?

Topic Test 3: Answers

1. **False.** For DNA polymerases to synthesize new DNA, the DNA templates need to be prepared. The process of template preparation involves four steps: unwinding of the double helix, breakage of the hydrogen bonds holding the two DNA strands together, coating of the single-stranded template strands to prevent them from reannealing, and primer synthesis. The enzymes responsible for unwinding and strand separation are called helicases and they recognize specific sequences known as origins of replication or *ori*.

2. **True.** Both *E. coli* DNA polymerases I and III have the ability to remove misincorporated nucleotides. Because DNA synthesis occurs in the 5′ to 3′ direction, the DNA polymerases need to remove nucleotides in the 3′ to 5′ direction. DNA polymerase I also has a 5′ to 3′ exonuclease activity, which it uses to remove the RNA primers.

3. **True.** Most eukaryotic chromosomes are much bigger than bacterial chromosomes. The large size of eukaryotic chromosomes requires that replication is initiated simultaneously at many different points along the chromosome to ensure that the DNA is fully replicated before cell division occurs.

4. **False.** DNA nucleotides can only add on to the 3′ end of a growing DNA chain. Because DNA strands are antiparallel, meaning that the 5′ end of one of the DNA strands is opposite the 3′ end of the other DNA strand, it follows that DNA polymerase must add

complementary bases in the 5′ to 3′ direction opposite a parental template strand that is oriented in the 3′ to 5′ direction. For example, consider the following template DNA strand:

3′ GGCATATTCGCTGCAGT 5′

The newly synthesized, antiparallel strand would be as follows:

5′ CCGTATAAGCGACGTCA 3′

Now, comparing the two strands in the same direction (both 5′ to 3′), it is apparent that they are not identical:

5′ TGACGTCGCTTATACGG 3′ (parental strand)

5′ CCGTATAAGCGACGTCA 3′ (new strand)

5. **d. Remove RNA primers.** Exonucleases remove nucleotides from an existing strand of DNA or RNA. Both DNA and RNA are synthesized in the 5′ to 3′ direction. Therefore, a DNA polymerase with 5′ to 3′ exonuclease activity would remove nucleotides from the "beginning" of a DNA or RNA strand (the 5′ end) rather than the "growing" end of the strand (the 3′ end). An important example of 5′ to 3′ exonuclease activity in DNA replication is the removal of RNA primers. The primers are required for DNA synthesis to occur because they provide a free 3′-OH group that DNA polymerases need to start adding nucleotides. However, the RNA primers must be removed and replaced with DNA, or a hybrid RNA–DNA strand would result.
6. **b. Purine–pyrimidine base pairs.** The pentose sugars and phosphate groups comprise the "backbone" of a DNA chain; in this analogy, they represent the handles of the ladder rather than the steps. a and c are therefore incorrect (c is additionally incorrect because ribose is the sugar found in RNA, not DNA). d and e are incorrect because complementary base pairing always matches adenines with thymines and guanines with cytosines.
7. **c. Synthesis of short RNA primers.** a is incorrect because Okazaki fragments are not removed, only the RNA primers used to start them. Furthermore, in *E. coli*, it is DNA polymerase I that removes these primers, not DNA polymerase III. Synthesis of DNA is continuous (no Okazaki fragments) on the leading strand, so b is incorrect. Each strand of a double helix is used as a template for replication and each template strand remains base paired with its newly synthesized complementary strand to create a new double-stranded daughter helix. Therefore, the parental double helix is not degraded as suggested in d. Finally, helicase, not DNA polymerase I, is responsible for unwinding the DNA double helix to allow replication to occur.
8. Semiconservative replication refers to the fact that after replication, both of the new double-stranded daughter helices are composed of one "old" DNA strand that originates from the original parental helix and one strand of completely newly synthesized DNA.
9. **DNA Ligase.** DNA ligase is the enzyme responsible for "sealing nicks" in newly synthesized DNA by catalyzing the formation of covalent bonds between the 3′-OH and 5′ phosphate groups of adjacent Okazaki fragments. In the absence of DNA ligase, the fragments produced on the lagging strand would not be joined together.

IN THE CLINIC

Gene therapy involves the direct delivery of a therapeutic gene into the cells of a particular tissue. An ideal gene delivery system would result in high efficiency of uptake of the therapeutic gene by the target cells, transportation of the therapeutic gene to the nucleus of the target cell with a minimum of intracellular degradation, and sustained expression of the therapeutic gene at a level that alleviates the condition being treated. Such an ideal gene delivery system might be possible with the construction of artificial chromosomes. An artificial chromosome could carry a large amount of DNA (e.g., a therapeutic gene plus all its regulatory sequences), and this DNA could be the original genomic version of the DNA. Both the long-term stability and sustained expression of a therapeutic gene should also be ensured due to the self-replicating nature of the artificial chromosome.

A detailed knowledge of each essential chromosomal element and chromosomal function, in particular the process of DNA replication, allowed researchers to construct the first artificial chromosomes. These artificial human chromosomes could be maintained in culture so the prospect of creating artificial human chromosomes carrying therapeutic genes is feasible. Although there are some drawbacks to this technology, such as the difficulty of delivering such a large DNA molecule to the nucleus of the target cell, artificial human chromosomes represent an exciting new tool for gene therapy.

DEMONSTRATION PROBLEM 1

Question

How did Meselson and Stahl determine that DNA replication is semiconservative rather than conservative or dispersive?

Solution

There are three hypothetical methods by which DNA can replicate itself. In the first, called **conservative**, the parental double helix provides the template, but the parental strands reassociate at the end of replication and the daughter strands form a double helix of their own. In this way, the parental double helix is always "conserved." A model of conservative replication is shown in the following figure (black lines represent parental DNA, gray lines represent newly synthesized DNA):

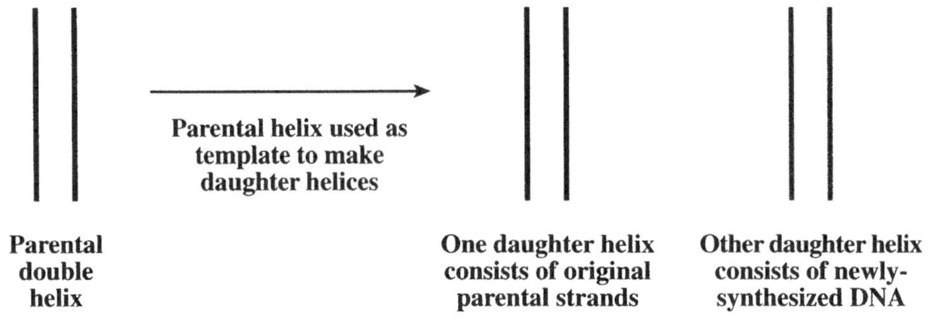

The second hypothetical replication method is called **semiconservative**. In semiconservative replication, each parental strand is used as a template for new DNA synthesis and remains base paired with that new strand. Each daughter helix is therefore a hybrid of one intact parental strand and one intact new strand. A model of semiconservative replication is shown in the figure below:

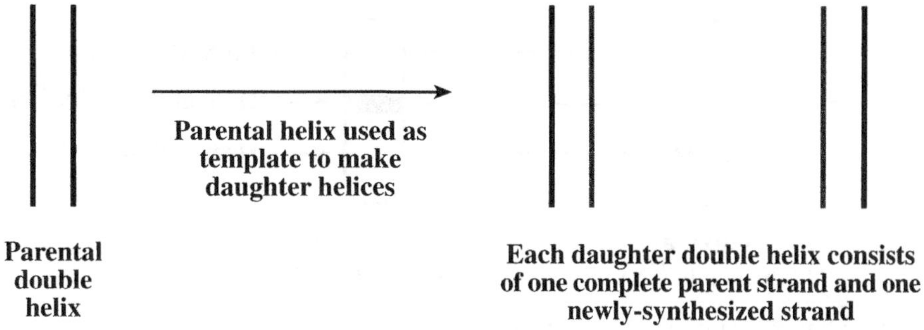

The third model is known as **dispersive** replication. The dispersive replication model suggests that fragments of DNA are replicated from either parental strand, resulting in double helices with parental and new DNA interspersed along each DNA strand (in contrast to the situation in semi-conservative replication in which the individual daughter strands of the "hybrid" double helices do not consist of a mixture of parental and new DNA). This model is shown in the following figure:

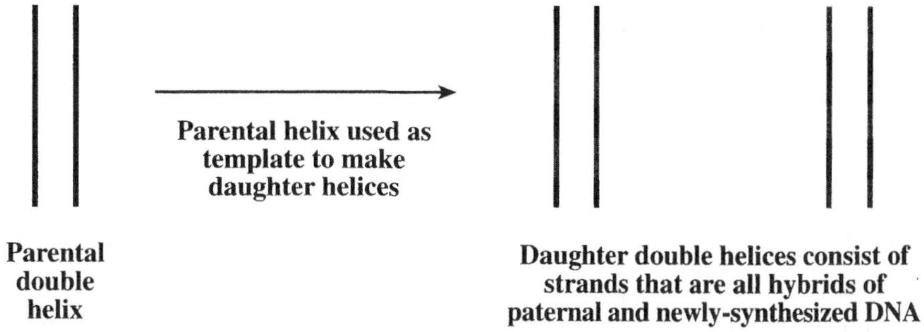

How then was it determined which one of these models is correct? If cultures of *E. coli* cells are grown in the presence of a radioactive isotope of nitrogen (^{15}N) rather than the isotope normally found in cells (^{14}N), when the nitrogen in the DNA is replaced with the "heavy" isotope, the cell's density increases enough that heavy DNA (DNA whose nitrogen content is entirely ^{15}N) and "light" DNA (all nitrogen is ^{14}N) can be separated and distinguished in a density gradient analysis experiment. When density analysis is performed on heavy and light cell cultures in control experiments, results are obtained that are indicated in the following figure:

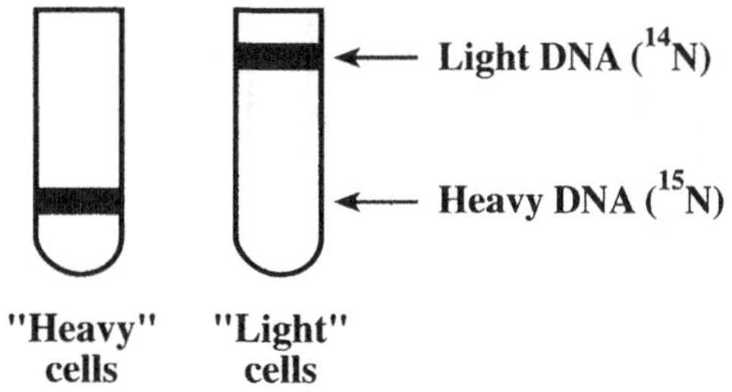

Meselson and Stahl grew cultures of *E. coli* in ^{15}N until all of the ^{14}N in their DNA had been replaced by ^{15}N. These cultures were then switched to media containing the light isotope of nitrogen, ^{14}N. The cells underwent one round of DNA replication in the light medium and then density gradient analysis was performed. If DNA replication is **conservative**, the parental (heavy) double helix would be reformed, whereas the newly synthesized "light" DNA would form its own double helix. The predicted density gradient would be one band for the heavy (parental)

double helix and another band for the light double helix. If DNA replication is either **semiconservative** or **dispersive**, all daughter helices would be hybrids of parental and newly synthesized DNA. In both of these cases, the predicted density gradient would be a single band of intermediate distance between the light and heavy bands from the control experiment. Meselson and Stahl's actual results are summarized in the following figure:

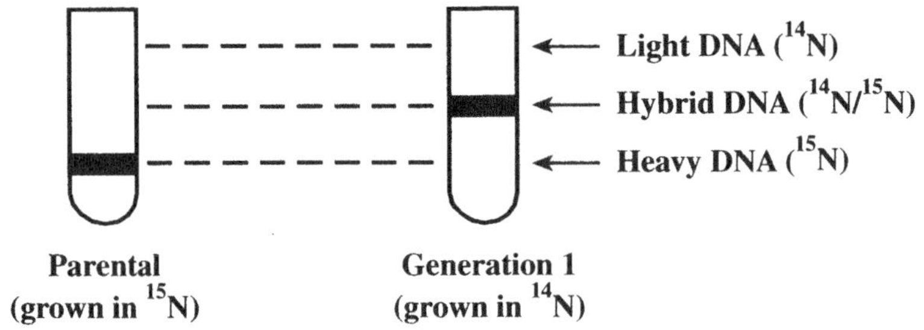

As the results indicate, DNA replication is not conservative, because the parental double helix is not maintained (the band corresponding to heavy DNA no longer appears). However, after one round of replication, Meselson and Stahl could not distinguish between the semiconservative and dispersive models. Therefore, Meselson and Stahl continued the experiment for a second round of replication in the light medium and again performed density gradient analysis of the resulting cultures. If DNA replication is dispersive, all daughter strands would be hybrids of parental and newly synthesized DNA and migrate to a point intermediate between light and heavy after one round. A second round of replication in the presence of ^{14}N would result in further hybridization of the DNA strands but with increasing light DNA content. The expected density gradient result would therefore be a single band migrating to a distance between light and heavy, but closer to light than the first generation.

On the other hand, if DNA replication is semiconservative, each daughter double helix would be a hybrid containing one intact parental strand and one intact newly synthesized strand. After one round of replication under these experimental conditions, this would result in each of the two daughter helices consisting of one heavy (parental) strand and one light (new) strand. During the second round of replication, each double helix would separate and each strand would be used as a template for the synthesis of new light DNA. As shown in the figure below, this would lead to each of the four parent strands being paired with a new light strand.

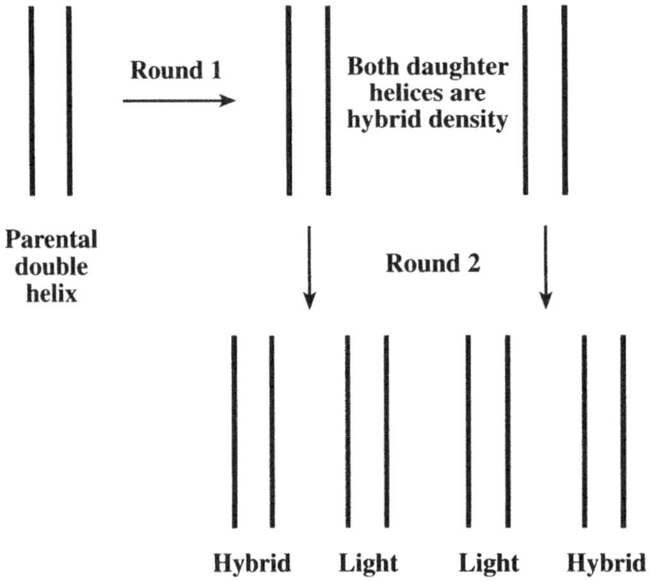

As evident from the figure, 50% of the resulting second-generation helices would still be of hybrid density, but the other 50% would be light. Because this is different than the expected result from the dispersive model, density gradient results after two rounds of replication in the presence of ^{14}N provided Meselson and Stahl with a method to eliminate one of the remaining two models. Their results are summarized in the following figure:

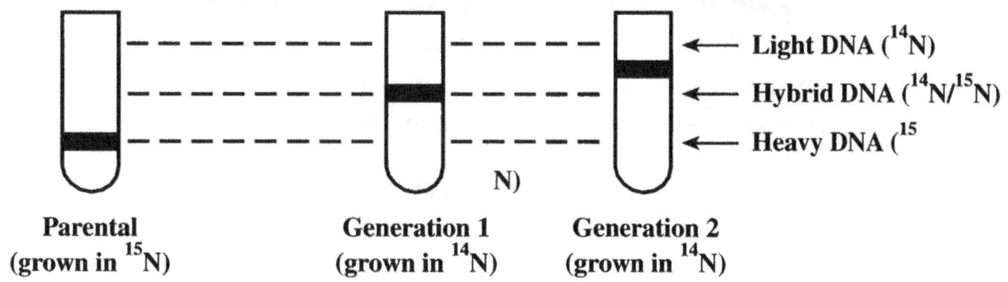

The presence of a light band accounting for 50% of the DNA indicated to Meselson and Stahl that DNA replication is semiconservative.

DEMONSTRATION PROBLEM 2

E. coli cells were grown in the presence of a heavy isotope of nitrogen (^{15}N) until all the normal nitrogen (^{14}N) in the DNA had been replaced by ^{15}N. The *E. coli* cells were then transferred into fresh medium containing normal (light) ^{14}N.

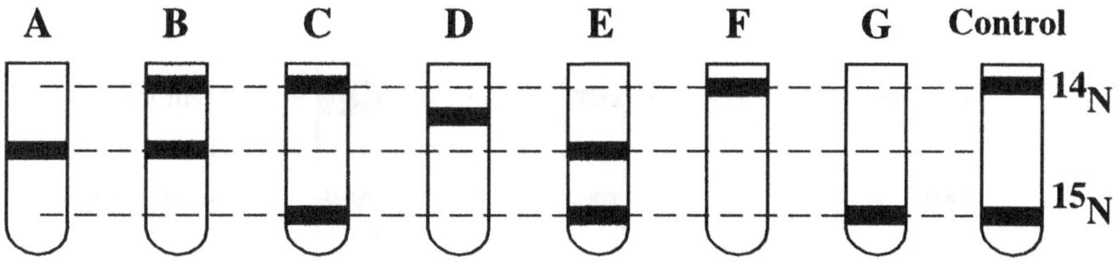

Which of the tubes shown above represents what Meselson and Stahl would have observed if DNA replication was **conservative** . . .

1. . . . After one round of replication?

The conservative DNA replication model predicts that the original parental double helix reforms after being used as a template for DNA synthesis, the newly synthesized strands forming a double helix that includes NO parental DNA. If this model was correct, after one round of replication, one of the double helices would consist of parental DNA that would be heavy (because the parental cells were grown in the presence of the heavy isotope of nitrogen) and the other would consist entirely of new DNA that would be light (because the medium in which the new DNA was synthesized contained the normal light isotope of nitrogen). This situation is represented by tube C and in the figure below.

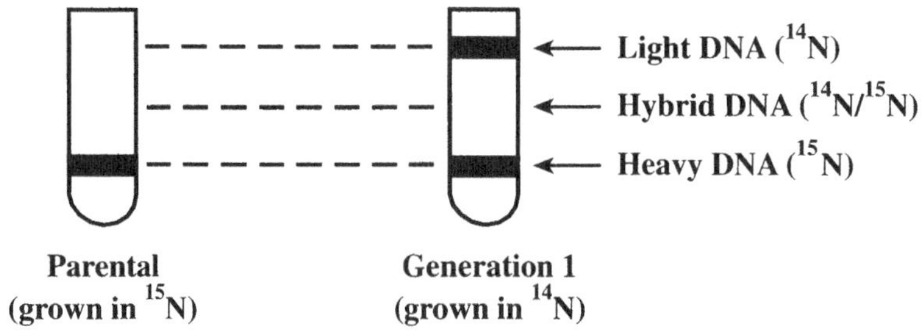

If DNA replication is CONSERVATIVE

2. . . . After two rounds of replication?

The only difference between this scenario and the scenario previously described for one round of replication is that we are now starting with two parental double helices, one heavy and one light. As before, the heavy parental double helix would yield two daughter double helices, one consisting entirely of the original parental DNA (heavy) and one consisting entirely of newly synthesized DNA (light). The light parental double helix would also give rise to two daughter double helices, one consisting of parental DNA (which was light in this case) and the other one consisting of newly synthesized DNA (also light). Therefore, after two rounds of replication, there would be four double helices, one of which would be heavy and three of which would be light. Tube C is again the correct answer, although the band migrating to the light position would be thicker than the corresponding band after one round of replication because the mass of DNA in that band would be greater, as demonstrated in the following figure.

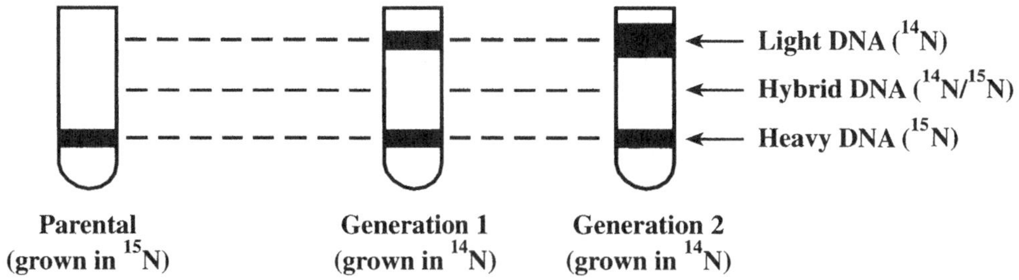

If DNA replication is CONSERVATIVE

Which of the tubes shown above illustrates what Meselson and Stahl would have observed if DNA replication was **dispersive** . . .

1. . . . After one round of replication?

The dispersive replication model suggests that fragments of DNA are replicated from either parental strand, resulting in double helices with parental and new DNA interspersed along each DNA strand. Therefore, after one round of replication, each of the daughter double helices would contain a mixture of heavy (parental) and light (newly synthesized) DNA in a 1 : 1 ratio.

These hybrid helices would migrate to a position approximately midway between heavy and light. This scenario is represented by tube A and shown in the figure below.

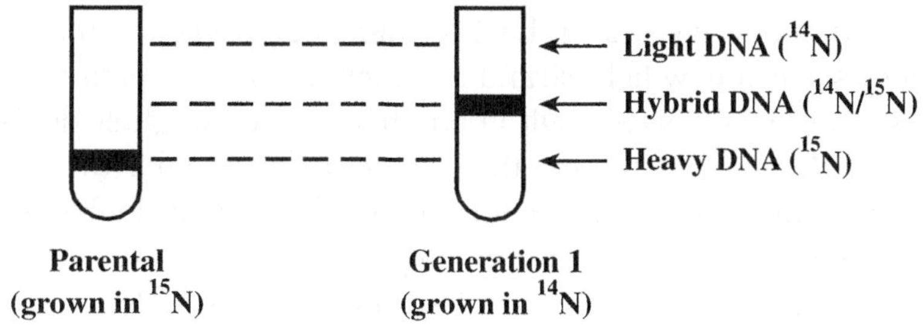

If DNA replication is DISPERSIVE

2. . . . After two rounds of replication?

The second round of replication would start out with two helices, both consisting of an equal mixture of heavy and light DNA. Each of these helices would replicate as before, resulting in a total of four helices that would consist of some heavy (parental) DNA and some light (newly synthesized) DNA. These new hybrid helices would migrate to a position somewhere between the midway and light bands because the amount of heavy DNA would now represent only 25% of the total DNA present. This scenario is represented by tube D and in the following figure.

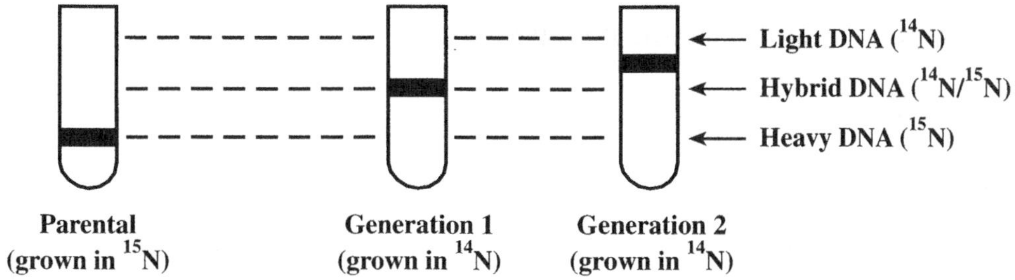

If DNA replication is DISPERSIVE

Which of the tubes shown above represents what Meselson and Stahl would have observed if DNA replication was **semiconservative** . . .

1. . . . After one round of replication?

In semiconservative replication, each parental strand is used as a template for new DNA synthesis and remains base paired with that new strand. Each daughter helix is therefore a hybrid of one intact parental strand and one intact new strand (this is in contrast to the "hybrid" helices described above where each daughter strand is itself a mixture of heavy and light DNA). Therefore, after one round of replication, the parental heavy helix would give rise to two hybrid daughter helices that would run midway between the heavy and light bands, as shown in tube A and in the figure below.

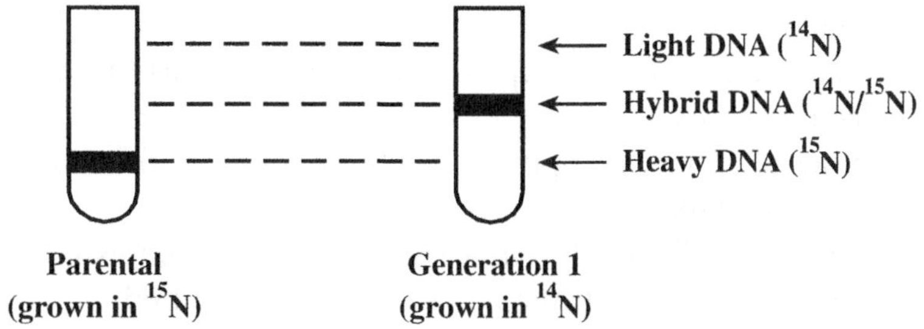

If DNA replication is SEMICONSERVATIVE

2. . . . After two rounds of replication?

The second round of replication would start out with two identical helices, each consisting of one light strand and one heavy strand. Each strand would separate and would be used as a template for the synthesis of a new light strand to which it would remain base paired. Thus, the two parental heavy strands would each result in another hybrid daughter double helix, consisting of one heavy strand and one light strand—the same situation as after the first round. The two parental light strands would result in the formation of two light double helices. So two bands would appear in the gradient, one corresponding to light double-stranded DNA molecules and one corresponding to hybrid ($^{14}N/^{15}N$) double-stranded DNA molecules, the situation shown in tube B and in the following figure.

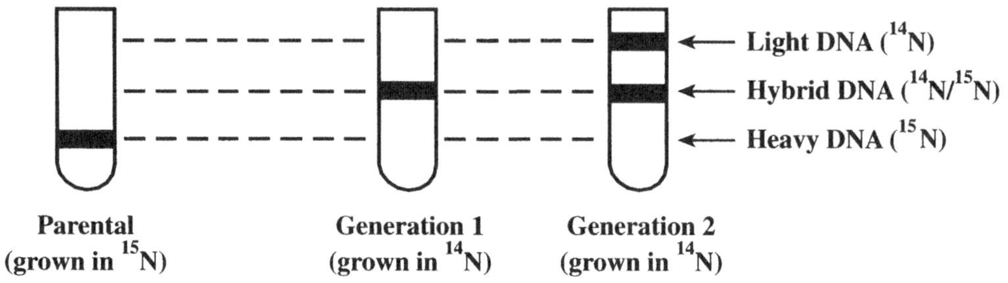

If DNA replication is SEMICONSERVATIVE

Chapter Test

True/False

1. The statement that the DNA double helix is "antiparallel" means that the nucleotides on one of the DNA strands are "upside-down" relative to the nucleotides on the opposite DNA strand.
2. If *E. coli* DNA polymerase I was affected by a mutation that eliminated its 3′ to 5′ exonuclease activity, synthesis of the Okazaki fragments would still occur.
3. Under normal denaturing conditions, the covalent phosphodiester bonds that connect the nucleotides in DNA chains remain intact.

Multiple Choice

4. Which one of the following *true* statements was **not** suggested by the rules formulated by Chargaff?
   a. The number of adenines is approximately the same as the number of thymines in any molecule of DNA.
   b. The number of guanines is approximately the same as the number of cytosines in any molecule of DNA.
   c. The number of purines is approximately the same as the number of pyrimidines in any molecule of DNA.
   d. If the percentage of adenines in a DNA molecule is known, the percentage composition of the other three bases can be determined.
   e. If the sequence of one DNA strand is known, the sequence of the other DNA strand can be determined.

5. Identify each of the lettered items in the following figure depicting DNA replication in *E. coli*:

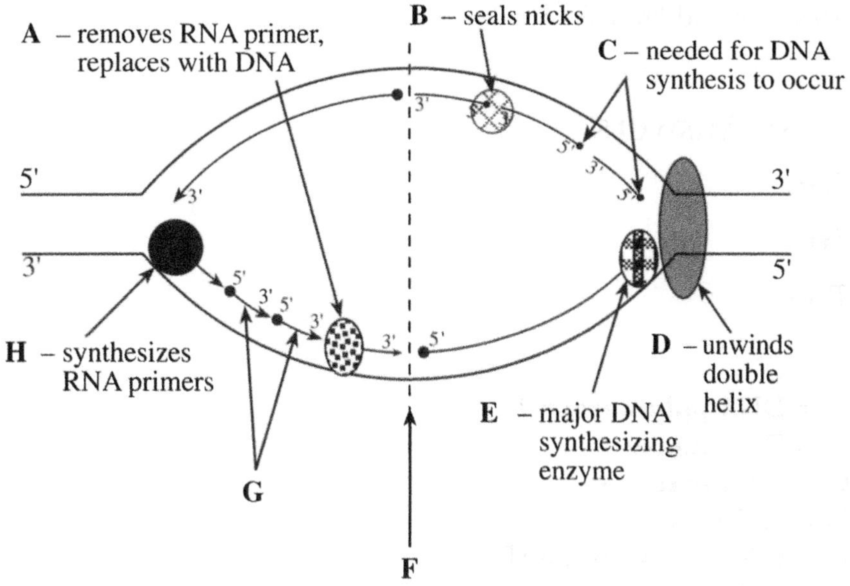

6. At the completion of DNA replication, each newly synthesized DNA strand is
   a. identical in sequence to the strand opposite which it was synthesized.
   b. complementary in sequence to the strand opposite which it was synthesized.
   c. a hybrid strand, consisting of both DNA and RNA nucleotides.
   d. oriented in the same direction as the strand opposite which it was synthesized.
   e. fragmented, consisting of multiple short DNA fragments.

Short Answer

7. The Meselson and Stahl experiment was used to determine the mode of replication of DNA in an organism from another planet. The organism was grown in the presence of a heavy isotope of nitrogen (^{15}N) until all the nitrogen in the DNA had been replaced by the heavy isotope. The organism was then transferred into medium containing the normal light isotope of nitrogen (^{14}N). At the conclusion of the experiment, it was announced that the alien organism used the **conservative** mode of replication. What were the results of the density gradient analysis after one round of replication in the light medium?

8. The sequence of one strand of DNA is 5′-AGTCGACGA-3′. What would be the 5′ to 3′ sequence of the complementary strand?

9. Four samples of double-stranded DNA are analyzed and the following data obtained:

Sample 1	30% thymine
Sample 2	15% cytosine
Sample 3	20% guanine
Sample 4	25% adenine

Which of these samples could represent DNA from identical sources?

Essay

10. Briefly outline the contributions made by Frederick Griffith, Oswald Avery, MacLyn McCarty and Colin MacLeod, Alfred Hershey and Martha Chase, and Matthew Meselson and Franklin Stahl.

Chapter Test Answers

1. **True**
2. **True**
3. **True**
4. **e**
5. **A = DNA polymerase I**
   B = DNA ligase
   C = RNA primers
   D = Helicase
   E = DNA polymerase III
   F = Origin of replication
   G = Okazaki fragments
   H = Primase (synthesizes RNA primers that initiate synthesis of the Okazaki fragments). RNA polymerase synthesizes the RNA primers that initiate synthesis of the leading strands.
6. **b**
7. **Two** distinct bands would appear, one corresponding to light DNA (the daughter double helix consisting entirely of newly synthesized DNA) and the other corresponding to heavy DNA (the original parental double helix containing the heavy isotope of nitrogen). The light DNA would band close to the top of the density gradient, whereas the heavy DNA would band close to the bottom of the density gradient.
8. **5′-TCGTCGACT-3′**
9. **Samples 1 and 3**
10. Griffith initially showed that the genetic material from heat-killed virulent bacteria could be taken up by live *non*-virulent bacteria, making them virulent. Based on these findings, Avery, McCarty, and MacLeod tested the properties of this "transforming factor" and found it to be DNA, thereby indicating that DNA is the genetic material.

    Hershey and Chase provided further evidence that DNA is the genetic material rather than protein. They did this by labeling the protein coat of one batch of a bacterial phage with ^{35}S and the DNA of another batch of the same phage with ^{32}P, allowing the phages to infect bacterial cells, then separating the phages from the bacteria and looking to see what had happened to the radioactive DNA and radioactive protein. Knowing that the phages transfer their genetic material to the bacteria, if protein is the genetic material, ^{35}S would be found inside the bacteria, but if DNA is the genetic material, ^{32}P would be found inside the bacteria. They found ^{32}P inside the cell and ^{35}S outside the cell, indicating that DNA is the genetic material.

Meselson and Stahl showed that DNA replication is semiconservative by growing *E. coli* in the presence of a heavy isotope of nitrogen (^{15}N) until all the nitrogen in the DNA was heavy. They then transferred the cells to light media (containing ^{14}N) so that all new DNA synthesis would incorporate light nucleotides. After one round of replication, all the DNA double helices were of hybrid density, eliminating the conservative model. After two rounds of replication, half of the DNA molecules were light and half were intermediate, eliminating the dispersive model.

Check Your Performance

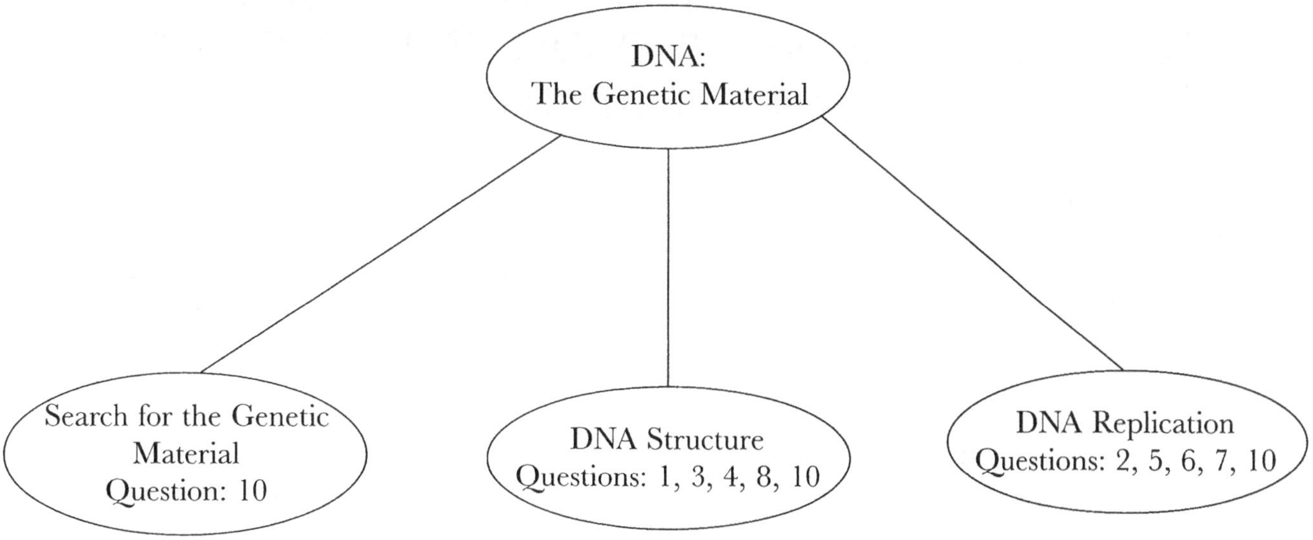

Note the number of questions in each grouping that you got wrong on the chapter test. Identify where you need further review and go back to relevant parts of this chapter.

CHAPTER 2

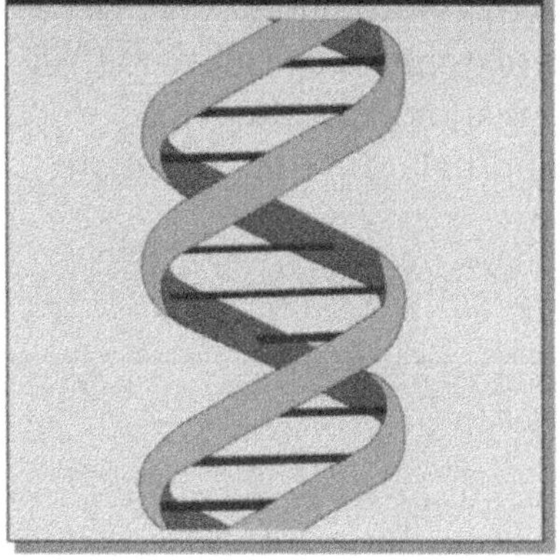

From DNA to RNA: The Process of Transcription

STUDY OF NUTRITIONAL MUTANTS OF THE COMMON BREAD MOLD PROVIDED EVIDENCE THAT GENES SPECIFY PROTEINS

Beadle and Tatum studied nutritional mutants of the common bread mold, *Neurospora crassa*. The wild-type mold can be grown on a very simple medium (a so-called minimal medium), containing little more than a carbon source, salts, and a few vitamins, but the nutritional mutants had lost the ability to survive on this simple medium. By growing the nutritional mutants on complete growth medium and then transferring cells to samples of minimal medium each supplemented with a different nutrient, Beadle and Tatum were able to identify the specific metabolic defect for each mutant. For example, they found three different classes of mutants all unable to synthesize arginine. By supplementing the medium with different intermediates, they showed that each mutant was blocked at a different step in the arginine biosynthetic pathway due to the loss of a single enzyme. Beadle and Tatum formulated the one gene–one enzyme hypothesis that states that the function of a gene is to control the production of a single enzyme. Beadle and Tatum's one gene–one enzyme hypothesis was later changed to the one gene–one protein hypothesis to acknowledge the fact that not all proteins are enzymes. Later still, this was changed to the one gene–one polypeptide hypothesis to account for the observation that many proteins are composed of more than one polypeptide chain, each of which is encoded by its own gene.

ESSENTIAL BACKGROUND

- **Structure of DNA**
- **Structure of RNA**

TOPIC 1: THE CENTRAL DOGMA

KEY POINTS

✓ *What do genes do?*

✓ *How do genes give rise to RNAs and proteins?*

✓ *Why does the use of an RNA intermediate enable cells to synthesize the required amount of protein much more rapidly than if DNA itself was used as the direct template for protein synthesis?*

Protein-coding genes do not control the production of proteins directly. Instead, the nucleotide information in DNA is first copied, by the process of **transcription**, into an RNA intermediate, called **messenger RNA** (**mRNA**). The nucleotide information in this mRNA is then "**translated**" into the amino acid sequence of proteins. This flow of genetic information, from DNA through an RNA intermediate into protein, is referred to as the **central dogma** of molecular biology.

Many identical RNA transcripts can be made from a single gene and each of these transcripts can be used to make many identical protein molecules. Because most genes are present in only one (bacteria) or two (eukaryotes) copies in a cell, the amplification in number of template molecules that occurs on passing from DNA to RNA enables cells to synthesize the required amount of protein much more rapidly than if DNA itself was used as the direct template for protein synthesis.

It is now known that some genes do not encode proteins but instead are transcribed into functional RNAs that play important roles in protein synthesis. In the case of these RNA-coding genes, the flow of genetic information stops at RNA.

Topic Test 1: The Central Dogma

True/False

1. In the central dogma, protein-coding genes are first copied into RNA intermediates, called ribosomal RNAs (rRNA).
2. Beadle and Tatum's one gene–one enzyme hypothesis was renamed one gene–one protein and then finally one gene–one polypeptide.

Multiple Choice

3. Transcription is the process of
   a. synthesizing a DNA molecule from an RNA template.
   b. using a DNA strand as a template to synthesize a complementary RNA molecule.
   c. using a DNA strand as a template to synthesize an identical RNA molecule.
   d. assembling ribonucleoside triphosphates (NTPs) into an RNA molecule without a template.
   e. synthesizing a protein using information carried in the nucleotide sequence of a messenger RNA.
4. Which of the following statements about Beadle and Tatum's experiments is correct?
   a. Their wild-type mold cultures were unable to grow on minimal media, but the nutritional mutants were capable of growing on minimal media.
   b. They hypothesized that each mutant was deficient in a single biosynthetic pathway and that each enzyme in the pathway was specified by the same gene.
   c. They observed that a mutation in a single gene destroyed the ability of that gene to synthesize a single enzyme.
   d. Each class of mutant defective in arginine synthesis could be supplemented with a common intermediate, demonstrating that a single gene leads to a single enzyme for each pathway.
   e. By supplementing the mold cultures with different proteins, they proved that one protein gives rise to one gene.

Short Answer

5. For protein-coding genes, the flow of genetic information is from ______ to ______ to ______.

Topic Test 1: Answers

1. **False.** The RNA species used as the intermediate is mRNA rather than rRNA.
2. **True.** This was done to emphasize that not all proteins are enzymes and also that many functional proteins are composed of multiple protein (polypeptide) chains.
3. **b.** e is not an incorrect statement; however, it describes translation rather than transcription. Transcription is the synthesis of an RNA molecule from a DNA template, eliminating d (transcription requires a template) and a, which is the reverse of transcription. Between options b and c, b is correct because the new RNA molecule is synthesized by adding NTPs that are complementary to the DNA template. Identical bases could not be added because they would not hydrogen bond with the template molecule.
4. **c.** Minimal media does not contain sufficient nutrients to support cultures that cannot synthesize all their own proteins. So a is incorrect because it is the wild-type cultures that are capable of growing on minimal media, not the nutritional mutants. b is not correct because each enzyme in a biosynthetic pathway is specified by its own gene. Likewise, a common intermediate cannot be used to supplement every mutant, because each class of mutants is missing a different intermediate. Finally, e is incorrect because Beadle and Tatum's hypothesis was that one gene gives rise to one protein, not the other way around.
5. DNA to messenger RNA to protein

TOPIC 2: COPYING DNA INTO RNA

KEY POINTS

✓ *What enzymes are responsible for copying DNA into RNA?*

✓ *Is the entire genome of an organism copied into RNA?*

✓ *How does RNA polymerase find genes?*

RNA POLYMERASES COPY DNA INTO RNA

RNA polymerases are the enzymes responsible for "transcribing" DNA into RNA. Most RNA polymerases are multisubunit enzymes, consisting of multiple polypeptide chains, and they are capable of locating and copying *gene* sequences in bacterial and eukaryotic genomes. The three stages of transcription are **initiation**, **elongation**, and **termination**.

Initiation

Eukaryotic and, to a lesser extent, bacterial genomes contain large amounts of noncoding DNA of no known function. This noncoding DNA is not copied into RNA—only coding DNA sequences or "genes" are "transcribed."

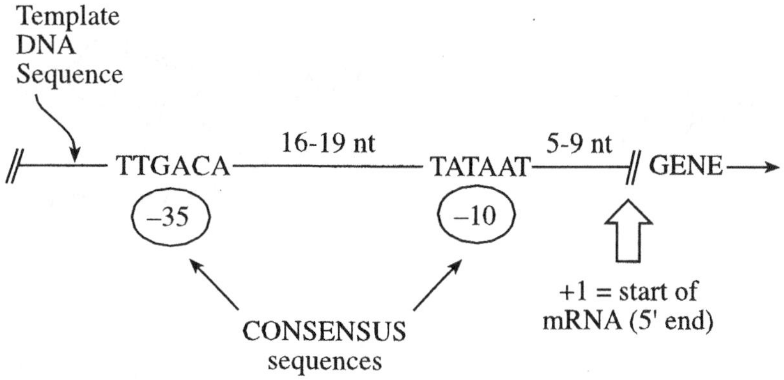

Figure 2.1 Bacterial consensus promoter sequences. The conserved sequences located "upstream" (or to the 5′ side of the transcription start site) of a typical bacterial gene are shown. RNA polymerase can recognize and bind to these promoter sequences, positioning itself so that it can begin synthesizing the RNA transcript at the first base in the gene, denoted above as "+1." One consensus sequence is found approximately 10 nucleotides (nt) upstream of the start of transcription and is called the "−10 site," whereas the other consensus sequence is found approximately 35 bases upstream of the +1 site and is called the "−35 site." (Sequences "after" or to the 3′ side of the transcription start site are said to be "downstream" of that gene.)

The expression of all genes is controlled by **promoter sequences**, two short (6–10 base pairs) highly conserved sequences located at specific distances from the start of the gene (**Figure 2.1**). In bacterial RNA polymerase, the so-called **sigma polypeptide** is responsible for recognizing these promoter sequences and positioning the RNA polymerase immediately in front of a gene sequence. In eukaryotes, additional proteins are required to help the RNA polymerases locate genes; these are discussed later in the chapter on genetic regulation in eukaryotes.

Once the RNA polymerase has been positioned on a promoter, it starts to unwind the DNA (**Figure 2.2**). Unwinding of the double helix is facilitated by the promoter sequence located closest to the start of the gene because this sequence is rich in adenine–thymine base pairs (Figure 2.1) that, by virtue of their lower number of hydrogen bonds, are easier to break than guanine–cytosine base pairs. Unwinding exposes DNA nucleotides that serve as a template for base pairing with RNA nucleotides.

Elongation

Using one of the DNA strands as a template, RNA polymerase links complementary RNA nucleotides that base pair along the template DNA strand. Unlike DNA polymerase, RNA polymerase does not require a primer to start RNA synthesis, but like DNA polymerase, RNA polymerase links RNA nucleotides to the 3′ end only of the growing chain. The 5′ to 3′ directionality of RNA synthesis means that RNA polymerase can only copy DNA strands that run in the opposite (3′ to 5′) direction. This observation, coupled with the fact that the promoter is asymmetrical and binds RNA polymerase in only one orientation, determines which DNA strand is used as the template. Once oriented by the promoter, RNA polymerase can only copy the sequence "in front of it" that runs in the 3′ to 5′ direction. The DNA strand that is transcribed (whether the "top" DNA strand or the "bottom" DNA strand) can be different for different genes located on the same DNA molecule, the orientation of the consensus promoter sequences being used to point the RNA polymerase in the appropriate direction for each gene (**Figure 2.3**).

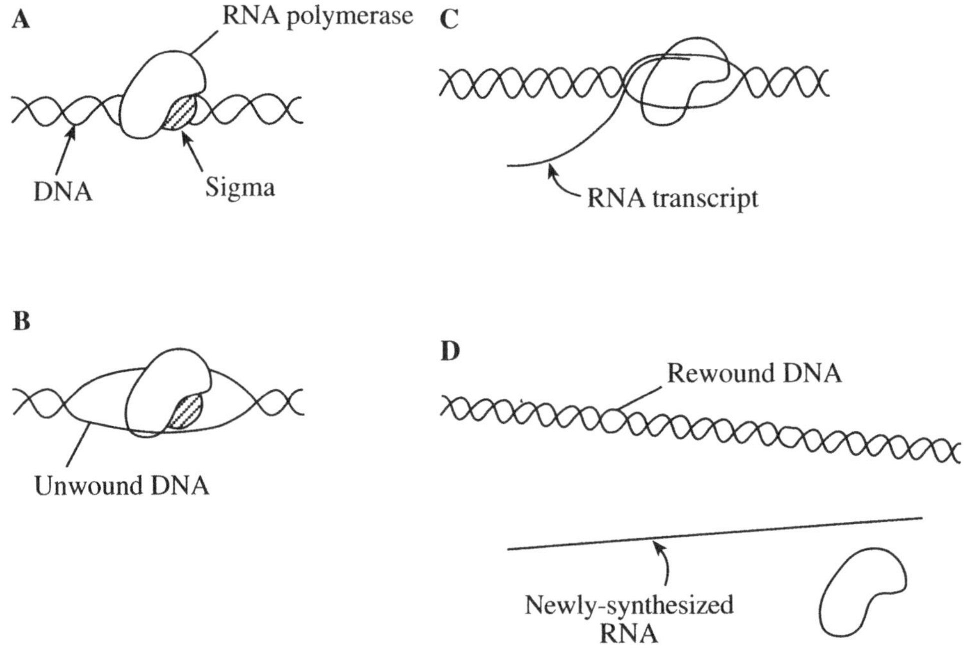

Figure 2.2 Steps in bacterial transcription. (A) Initiation: promoter recognition. Transcription begins when RNA polymerase locates the consensus promoter sequences on one of the strands of a DNA double helix. In bacteria, the sigma subunit of the RNA polymerase is responsible for locating promoter sequences; in eukaryotes additional proteins are required for efficient initiation of transcription. (B) Initiation: unwinding. Once the polymerase is positioned on the promoter, the DNA helix is unwound to expose the single strand of DNA that will be used as a template for RNA synthesis. (C) Elongation. Using the DNA strand as a template, RNA polymerase adds complementary RNA nucleotides in the 5′ to 3′ direction opposite the DNA bases. The growing RNA transcript does not remain base paired to the DNA template; instead the DNA double helix rewinds behind the advancing RNA polymerase, at the same time displacing the newly synthesized RNA that remains single stranded. (D) Termination. RNA synthesis continues until the RNA polymerase encounters a specific termination sequence in the template DNA. At this point, the RNA polymerase releases both the new RNA transcript and the DNA (which is now restored to its original double-helical structure).

The growing RNA chain does not remain base paired to the DNA template strand but instead is peeled away as the double helix rewinds behind the advancing RNA polymerase (Figure 2.2).

Termination

In both bacteria and eukaryotes, RNA synthesis continues along the DNA template until the RNA polymerase encounters specific base sequences in the DNA, called **terminators**. In bacteria, these sequences may be recognized by the polymerase itself (intrinsic or rho-independent termination) or by a special termination protein referred to as the Rho factor (rho-dependent termination). In intrinsic termination, the terminator consists of an inverted-repeat sequence followed by an adenine-rich region. When RNA polymerase transcribes the termination sequences, the inverted-repeat sequence causes intrastrand base pairing in the RNA, forming a hairpin loop structure, and the adenine-rich region is transcribed into a sequence of six to eight uracils. Base pairing of the RNA to itself means that less of the RNA transcript (only the short stretch of uracils) is available for base pairing to the DNA. This leads to dissociation of the completed

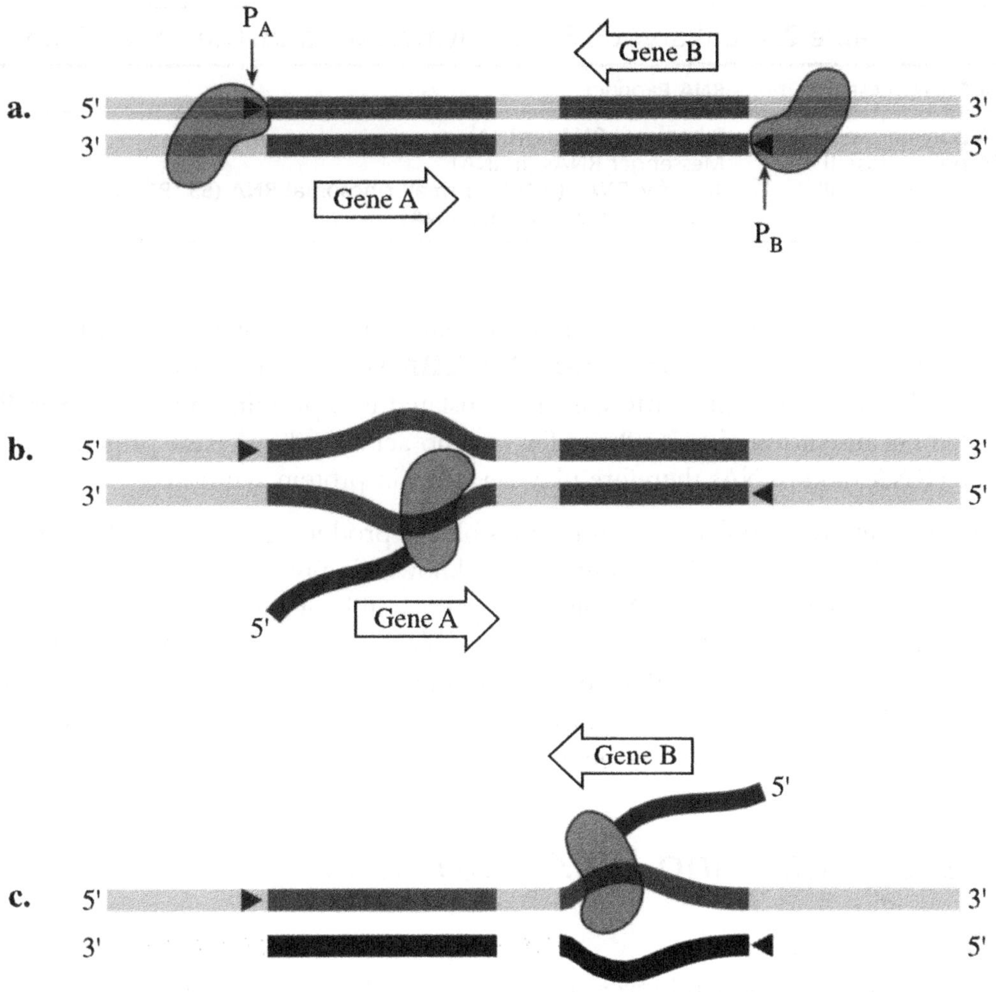

Figure 2.3 Directionality of transcription initiation. (A) Region of DNA with two genes, gene A and gene B. Each gene has a promoter sequence preceding it on the 5′ to 3′ strand. P_A designates the promoter for gene A; P_B is the promoter for gene B. The asymmetry of the promoter site ensures the correct orientation of RNA polymerase. (B) Transcription of gene A. RNA polymerase recognizes and binds P_A on the 5′ to 3′ strand but must use the 3′ to 5′ strand as the template to synthesize RNA in the 5′ to 3′ direction. This corresponds to the bottom strand in the figure. The thick black line represents the newly synthesized RNA transcript. (C) Transcription of gene B. RNA polymerase recognizes and binds P_B on the 5′ to 3′ strand but must use the 3′ to 5′ strand as the template to synthesize RNA in the 5′ to 3′ direction. This corresponds to the top strand in the figure. The thick black line again represents the newly synthesized RNA transcript.

RNA chain from the DNA. In rho-dependent termination, Rho binds to a special termination sequence in the RNA and forcibly pulls the RNA away from the DNA template.

Termination in eukaryotes is not well understood. As with bacteria, the terminators may be recognized by the polymerase or by as yet unidentified protein factors.

SEVERAL TYPES OF RNA ARE PRODUCED IN CELLS

Most genes specify the amino acid sequence of proteins, and the RNA molecules that RNA polymerase makes from these genes (and that ultimately direct the synthesis of proteins) are called **messenger RNAs**. However, two other major types of RNA are also made; the differ-

Table 2.1 Eukaryotic RNA Polymerases and Their RNA Products	
RNA Polymerase	**RNA Product**
RNA polymerase I	Ribosomal RNAs (rRNA)
RNA polymerase II	Messenger RNAs (mRNA)
RNA polymerase III	Transfer RNAs (tRNA), a small ribosomal RNA (5S rRNA), and some other small RNAs

ence between these RNAs and the mRNA is that, in the case of the non-mRNAs, the RNA itself is the final product. Thus, **ribosomal RNA (rRNA)** is so called because it is a major component of ribosomes, on which mRNAs are translated into proteins, and **transfer RNA (tRNA)** is so called because it functions to "transfer" amino acids to the sites of protein synthesis on ribosomes. rRNAs and tRNAs therefore play key roles in protein synthesis.

In bacteria, one RNA polymerase is responsible for producing all three RNA types. However, in eukaryotes, three different RNA polymerases, known simply as RNA polymerases I, II, and III, are required. Each of these RNA polymerases is capable of recognizing the promoter region specific to a particular gene/RNA type. Thus, in eukaryotes, genes that specify mRNAs are preceded by promoter sequences that differ from the promoter sequences found in front of genes that give rise to rRNAs and tRNAs. Likewise, the promoter sequences that precede genes for most rRNAs differ from those that precede the genes for tRNAs and some small rRNAs (**Table 2.1**).

Topic Test 2: Copying DNA into RNA

True/False

1. RNA polymerase synthesizes an RNA double helix from a DNA template.
2. In eukaryotes, one of the RNA polymerase enzymes is used to synthesize tRNAs and some small rRNA molecules.
3. When DNA is copied into RNA, the same strand of the DNA double helix is used as the template for the entire length of the chromosome.

Multiple Choice

4. What is the sequence of events during transcription?

I. Initiation and elongation of the mRNA strand.
II. Binding of an RNA polymerase to DNA.
III. Separation of the two strands of DNA.
IV. Release of the completed mRNA from the DNA.

a. III, II, IV, I
b. II, III, I, IV
c. III, I, II, IV
d. II, I, III, IV
e. III, II, I, IV

5. The direction of synthesis of a new mRNA molecule is

a. 5′ to 3′ from a 5′ to 3′ DNA template strand.

b. 5′ to 3′ from a 3′ to 5′ RNA template strand.
c. 5′ to 3′ from a 5′ to 3′ RNA template strand.
d. 5′ to 3′ from a 3′ to 5′ DNA template strand.
e. 3′ to 5′ from a 5′ to 3′ DNA template strand.

6. Which of the following statements about RNA polymerase is incorrect?
   a. RNA polymerase can initiate RNA synthesis without a DNA or RNA primer.
   b. Synthesis of RNA by RNA polymerase occurs in the 5′ to 3′ direction.
   c. RNA polymerase synthesizes an RNA transcript using ribonucleoside triphosphates (NTPs).
   d. Correct initiation of RNA polymerase in bacteria requires the sigma polypeptide.
   e. None of the above

Short Answer

7. ________ are highly conserved nucleotide sequences in the DNA that determine the sites where RNA synthesis is initiated.

8. The promoter sequence located closest to the start of a gene generally consists of just two of the four DNA nucleotides. What are these two nucleotides and why?

Topic Test 2: Answers

1. **False.** RNA polymerase synthesizes single-stranded RNA from a DNA template.

2. **True.** There is also a separate RNA polymerase enzyme in eukaryotes for mRNA molecules and a third enzyme for the remaining rRNA molecules.

3. **False.** The DNA strand that is transcribed (whether the "top" DNA strand of a double helix or the "bottom" DNA strand) can be different for different genes located on the same DNA molecule, the orientation of the consensus promoter sequences being used to point the RNA polymerase in the appropriate direction for each gene.

4. **b.** RNA polymerase must first find and bind to the promoter site preceding the gene (II). It then unwinds the DNA to single strands, one of which can be used as a template (III). This is followed by the synthesis of a complementary RNA strand (I), which ends when the RNA polymerase encounters a special termination sequence, causing it to release the completed RNA strand (IV).

5. **d.** Nucleic acid synthesis, RNA and DNA, always occurs in the 5′ to 3′ direction, which eliminates e. Synthesis of mRNA (in the process known as transcription) requires a DNA template, so b and c are incorrect. This leaves a and d, of which d is correct because RNA polymerase can only copy DNA strands that run in the opposite (3′ to 5′) direction.

6. **e.** All options given are correct in describing RNA polymerase. As opposed to DNA polymerase, which requires an RNA primer, RNA polymerase does not require a primer to initiate synthesis. But like DNA polymerase, it synthesizes its product (RNA) in the 5′ to 3′ direction using NTPs (DNA polymerase actually uses deoxyribonucleoside triphosphates [dNTPs]). In bacteria, the "sigma" subunit of the RNA polymerase enzyme is needed to locate the promoter sequences preceding the gene.

7. Promoters

8. The promoter sequence located closest to the start of genes is rich in adenine and thymine. This makes the DNA double helix easier to denature and unwind for transcription to occur.

TOPIC 3: RNA PROCESSING

KEY POINTS

✓ *Does all the noncoding DNA found in eukaryotic genomes occur between genes?*

✓ *Why are mature eukaryotic RNAs shorter than the genes from which they are copied?*

✓ *How are noncoding sequences that interrupt genes removed?*

EUKARYOTIC RNAs ARE MODIFIED BEFORE LEAVING THE NUCLEUS

Although the process by which RNAs are produced is similar in all organisms, what happens to the resulting RNA transcripts before they can be used by the cell differs considerably between eukaryotes and bacteria.

In bacteria there is no distinct nucleus so transcription and translation both occur in the cytoplasm. What this means is that transcription and translation can occur simultaneously in bacteria: As RNA polymerase is synthesizing a bacterial mRNA, ribosomes can attach to the free 5′ end of the incomplete mRNA and initiate translation. In eukaryotes, the processes of transcription and translation are clearly separated by a nuclear membrane. Transcription occurs in the nucleus and translation takes place on the surface of ribosomes in the cytoplasm. Therefore, before a eukaryotic RNA can be used by the cell, it first has to be transported out of the nucleus. Before eukaryotic RNAs can leave the nucleus, however, they have to undergo various degrees of RNA processing.

EUKARYOTIC GENES ARE INTERRUPTED BY NONCODING SEQUENCES

Eukaryotic genomes especially contain a lot of noncoding DNA. Some of this noncoding DNA occurs between genes, where it is referred to as **spacer DNA**, but noncoding DNA also occurs within most eukaryotic genes, in which case it is referred to as **intervening or intron DNA**. This surprising feature of eukaryotic genes is fundamentally different from the situation in bacterial genes which, with a few exceptions, are not interrupted by introns. The discovery of noncoding sequences that interrupt genes also explained the mysterious observation that mature eukaryotic RNAs are usually considerably shorter than the gene sequences from which they are derived because intron sequences can vary from 80 to 10,000 nucleotides in length and they have to be removed before the RNA transcripts can be used by the cell.

In eukaryotes, the RNA polymerases copy entire gene sequences into RNAs. Therefore, the nascent RNA transcripts still contain the intron sequences and are referred to as the **primary**

or initial transcripts. Special enzymes have to remove the introns and join together the remaining coding sequences (the so-called **exons**) to form a continuous uninterrupted copy of the gene before the RNA transcript can leave the nucleus. This process is known as **RNA splicing**, and it occurs in the production of all three types of eukaryotic RNA (mRNA, tRNA, and rRNA).

How does the cell know where an exon stops and an intron starts? The exact nucleotide sequence of most of an intron seems to be unimportant, and consequently there is little resemblance between the sequences of different introns. The only sequences that are the same or very similar in all introns are short nucleotide sequences at each end of an intron. These short conserved sequences are important because they differentiate intron from exon sequences and, in nuclear primary mRNA introns, they are recognized by **small nuclear ribonucleoproteins (snRNPs)**—pronounced "snurps"—complexes composed of proteins and **small nuclear RNA**. snRNPs assemble at the conserved nucleotide sequences at either end of an intron and excise it by cutting the DNA at each exon–intron junction. The first cut releases the intron as a "lariat" and the second cut results in the complete release of the intron. The cut RNA chain is then rejoined to form a continuous coding sequence (**Figure 2.4a**).

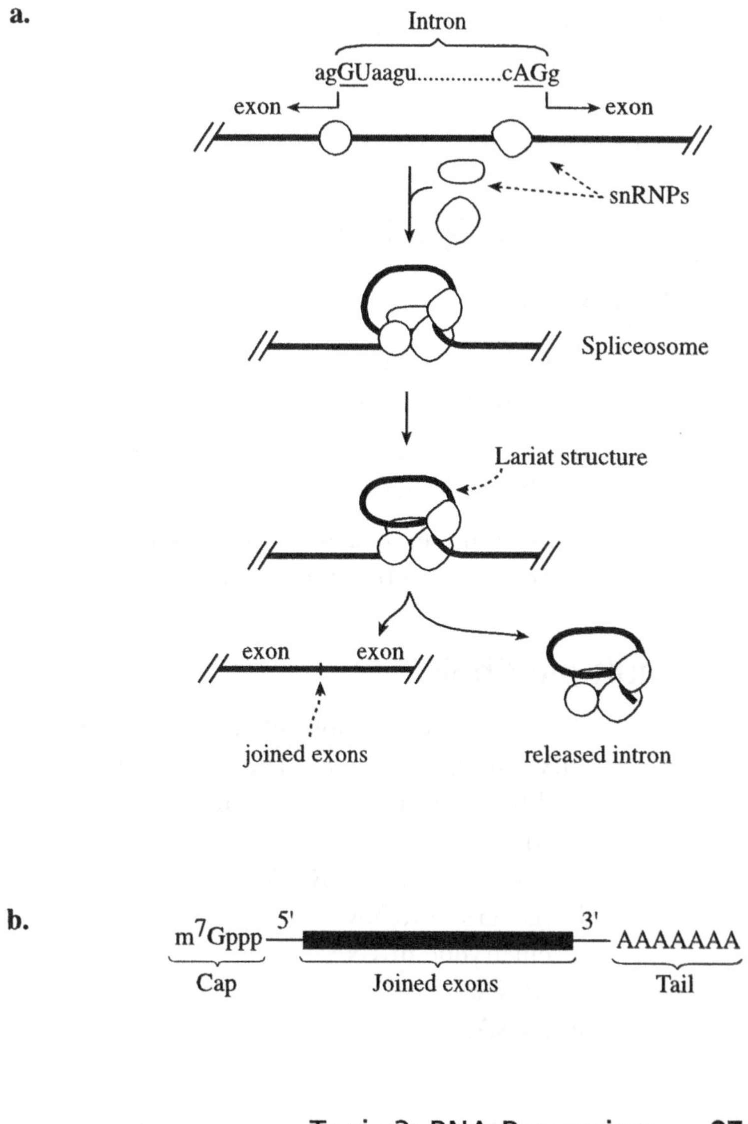

Figure 2.4 RNA processing in eukaryotes. (a) Intron removal from a primary mRNA transcript. First, snRNPs position themselves at, or close to, both ends of the intron. Other snRNPs then join to form an assembly called a **spliceosome** in which the ends of the intron are drawn together to form a looped structure. A cut is then made in the first intron–exon junction resulting in the formation of a **lariat** structure. A second cut releases the intron from the RNA transcript and the exons are spliced together to form a continuous coding sequence. (b) Mature mRNA structure. After intron removal, both ends of a mRNA have to be modified before the mRNA can pass from the nucleus into the cytoplasm. At the 5′ end, a methylguanosine cap structure is added, whereas at the 3′ end, a long "tail" of adenine residues (the so-called polyA tail) is added.

mRNA has to be further processed before it can leave the nucleus and be used to direct the synthesis of proteins. In **RNA capping**, a modified guanine nucleotide (a guanine with a methyl group attached) is added to the $5'$ end of all eukaryotic mRNAs. This $5'$ cap structure is recognized by part of the protein synthesizing apparatus, and it determines where protein synthesis starts on the mRNA. In **polyadenylation**, a long tail of adenine residues is added to the $3'$ end of eukaryotic (and some bacterial) mRNAs. The exact function of the $3'$ poly(A) tail is not known, but it may serve to protect the ends of mRNAs from degradation by hydrolytic enzymes.

Only after the introns have been removed and the $5'$ cap and $3'$ poly(A) tail have been added can eukaryotic mRNAs leave the nucleus and be used to direct protein synthesis in the cytoplasm (**Figure 2.4b**). Both tRNA and rRNA molecules are also produced from larger primary transcripts. Often a single transcript contains the sequences for several different rRNA molecules or for both tRNA and rRNA molecules. Formation of rRNA and tRNA molecules usually involves cleavage by several enzymes and chemical modification of various bases.

SOME RNA MOLECULES HAVE ENZYMATIC ACTIVITY

Some RNA molecules alone can catalyze particular reactions. These recently discovered catalytic RNAs are referred to as **ribozymes**. For example, RNase P is a widely occurring RNA molecule that acts as an endonuclease that processes primary tRNA transcripts, and yet another catalytic RNA molecule, discovered in *Tetrahymena*, removes intron sequences from primary rRNA transcripts. The removal of noncoding sequences by catalytic RNA molecules is referred to as **self-splicing** because it occurs in the absence of protein. (What does the discovery of ribozymes do to Beadle and Tatum's "one gene–one polypeptide" hypothesis?)

Topic Test 3: RNA Processing

True/False

1. snRNPs assemble at intron–exon junctions and release introns from the primary transcript after the primary transcript has entered the cytoplasm.
2. The $5'$ cap structure on a bacterial mRNA is used to locate the start site for protein synthesis on that mRNA.
3. Noncoding sequences are only found in, and removed from, primary RNA transcripts that are destined to become messenger RNAs.

Multiple Choice

4. A single-stranded molecule is synthesized from a DNA template according to complementary base pairing rules. The newly synthesized single-stranded molecule is then shortened and the shortened molecule used as a template for the synthesis of a protein. The shortened molecule is most likely
   a. primary or initial RNA.
   b. bacterial mRNA.
   c. eukaryotic mRNA.
   d. tRNA.
   e. rRNA.

5. Which of the following statements about introns is incorrect? Introns are
   a. found in most eukaryotic genes.
   b. removed during RNA processing.
   c. normally not transcribed.
   d. normally not translated.
   e. responsible for the fact that most eukaryotic mRNAs are much shorter than the genes from which they are derived.

6. Which of the following processes occur in the nucleus of a eukaryotic cell?
   a. DNA replication
   b. Transcription
   c. Translation
   d. RNA processing
   e. None of the above

Short Answer

7. Describe how pre-mRNA molecules are processed before they can be translated.

8. As an intron in a nuclear primary mRNA is removed, it is looped into a structure that is referred to as a ______.

Topic Test 3: Answers

1. **False.** snRNPs are responsible for intron splicing in eukaryotes, but they act while the primary transcript is still located in the nucleus, not after the mRNA has entered the cytoplasm.

2. **False.** The 5′ cap structure is used to locate the start site for protein synthesis, but on eukaryotic mRNAs, not bacterial mRNAs.

3. **False.** Noncoding sequences are also found in the primary transcripts for rRNAs and tRNAs.

4. **c.** Primary or initial RNA is the RNA transcribed from eukaryotic genes that contain introns. Primary RNA is the same length as the gene from which it is transcribed because RNA polymerase copies the entire DNA sequence (coding and noncoding) into RNA. The introns are then removed and the exons joined together, resulting in an RNA with a continuous coding sequence that is shorter than the original gene. Only mRNAs serve as templates for protein synthesis, suggesting b or c. Because most bacterial genes do not contain introns, c is the most likely answer.

5. **c.** All statements about introns are correct except for c. Transcription is the process of synthesizing RNA from DNA. In eukaryotes, the RNA polymerases transcribe *all* of the DNA sequence between the promoter and terminator sequences. Therefore, introns are normally transcribed. However, by the time most mRNA molecules are ready to be translated, the introns have been removed, so the introns are not translated.

6. **a, b, and d.** DNA replication occurs in the nucleus because this is where the genetic material is located. By the same reasoning, if DNA is used as a template for RNA synthesis, transcription must occur in the nucleus as well. RNA processing also takes place

in the nucleus, after which a mature mRNA molecule is transported into the cytoplasm for translation into protein.

7. Primary transcripts destined to become mRNAs have their introns removed by snRNPs. 3′ poly-adenine (3′ polyA) tail and 5′ methyl-guanosine cap structures are also added.
8. Lariat

IN THE CLINIC

In many cases, it is advantageous to block or decrease the production of certain proteins. This can be accomplished by antisense regulation. In antisense gene regulation, small nucleic acids are used to prevent translation of the information carried on a mRNA into a protein. For example, one of the first successful applications of antisense technology was in the production of Calgene's FlavR SavR Tomato. Grocery store tomatoes often lack flavor because they are picked green. Green tomatoes are hard and therefore easier to harvest and ship than soft ripe tomatoes. Scientists discovered that the softening of tomatoes during ripening is caused by the enzyme polygalacturonase. Blocking production of this enzyme would allow a tomato to ripen and hence develop flavor without becoming soft. To block production of polygalacturonase, a piece of RNA was made that was exactly complementary and opposite in polarity to the normal mRNA for the polygalacturonase protein. This antisense RNA hybridizes to the polygalacturonase mRNA, creating a double-stranded RNA molecule that cannot be translated into protein.

Many biotechnology companies view antisense technology as an exciting new tool for controlling gene expression in crop plants, fighting viral and bacterial diseases, and treating cancer and other human diseases. The rationale is simple: once the target RNA sequences have been identified, antisense RNAs (or DNAs) can be designed and delivered to the appropriate cells. The antisense RNA or DNA must be specific for the target sequence to prevent any undesirable side effects due to blocking of other proteins in the host cell. Further modifications of the antisense nucleic acids may be necessary to prevent them from being degraded in the cell but these complications are resolved, the potential applicability of this technology is enormous.

DEMONSTRATION PROBLEM

The following gene is isolated:

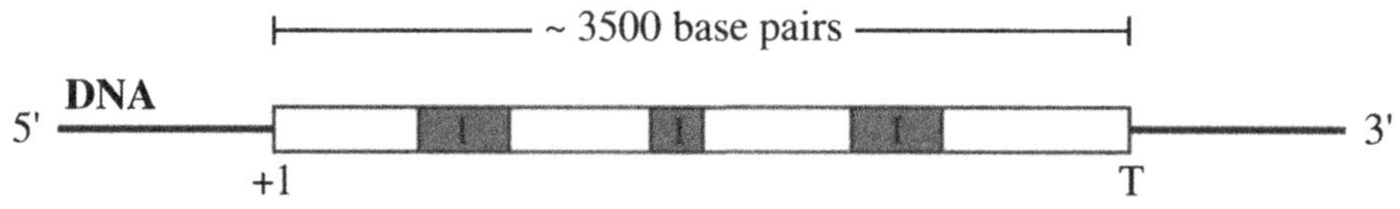

The gene is approximately 3,500 base pairs in length and is interrupted by three noncoding regions called introns (labeled "I"). "+1" indicates the first base that will be transcribed into RNA, and "T" denotes the terminator sequence that signals the end of transcription.

Question

Knowing nothing else about this gene other than the information already provided, would you say this gene was more likely to be a bacterial gene or a eukaryotic gene?

Solution

Bacterial genomes are smaller than eukaryotic genomes. One reason for this is that bacteria require fewer proteins to survive and reproduce. Another reason is that much of a eukaryote's genome consists of noncoding DNA and this noncoding DNA occurs between genes (so-called intergenic or spacer DNA) but also within genes (so-called intragenic or intron DNA). Bacterial genomes do not contain as much intergenic noncoding DNA, and most bacterial genes do not contain introns. For this reason, the gene shown above is most likely a eukaryotic gene.

Question

What would the primary RNA transcript of this gene look like?

Solution

RNA polymerase cannot distinguish coding from noncoding DNA. Its only responsibility is to recognize the promoter (and in some cases the terminator sequences), bind to the DNA, and use the DNA as a template to make a complementary strand of RNA. Therefore, the strand of RNA that is synthesized by RNA polymerase, called the **primary transcript**, will contain all the information found in the DNA located between the promoter and terminator sequences. This includes the intron DNA, which has to be removed later to generate a continuous protein-coding sequence. Therefore, the primary transcript is approximately the same length as the DNA gene sequence and would resemble the following figure (remember that the figure now shows RNA rather than DNA):

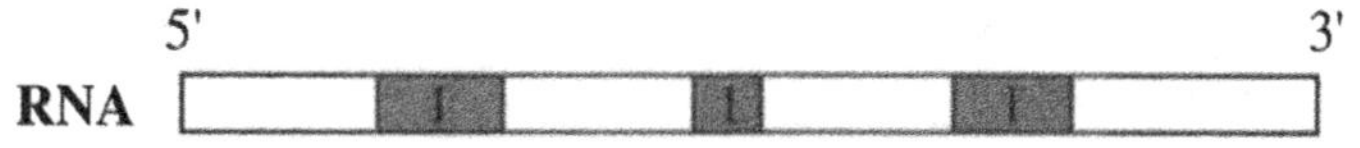

Primary transcript

All primary RNA transcripts (whether they are destined to become mRNAs, tRNAs, or rRNAs) are modified posttranscriptionally.

One of the processing steps of pre-mRNA transcripts involves intron removal. This is accomplished, as described earlier in the chapter, by snRNPs.

Question

What would the pre-mRNA molecule look like after intron removal?

Solution

During intron removal, the ends of the coding sequences (called **exons**) are spliced together to form a continuous protein-coding sequence. The process of intron removal needs to be very precise because the incorrect addition or removal of even a single base would very likely result in the synthesis of an inactive protein. Once the introns have been removed, the RNA transcript would be much shorter than the original DNA sequence and primary RNA transcript as depicted below:

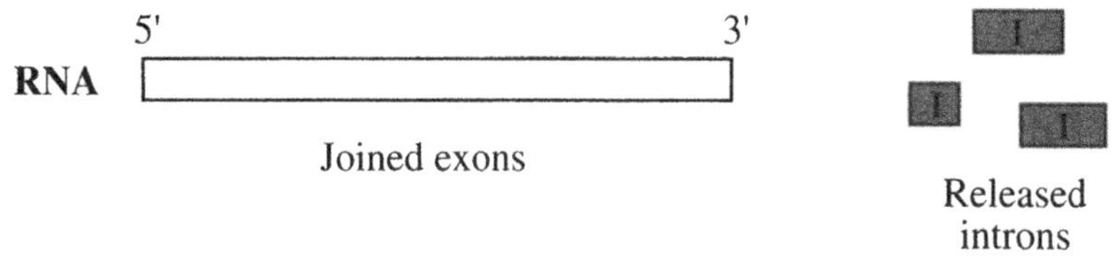

Finally, if this was a pre-mRNA molecule, both ends of the pre-mRNA transcript would also need to be modified.

Question

What further modifications need to occur to make this pre-mRNA transcript ready for export into the cytoplasm?

Solution

The 5′ end of the pre-mRNA transcript must have a "cap" structure attached to it. This cap is a modified methylated guanine nucleotide, and the structure is often referred to as a **7-methyl-guanosine cap**. At the 3′ end of the transcript, a long string of adenine nucleotides is added; this structure is called a **3′ poly(A) tail**. After these structures have been added to the pre-mRNA transcript, the resulting mature mRNA molecule can be exported into the cytoplasm to be used as a template for protein synthesis.

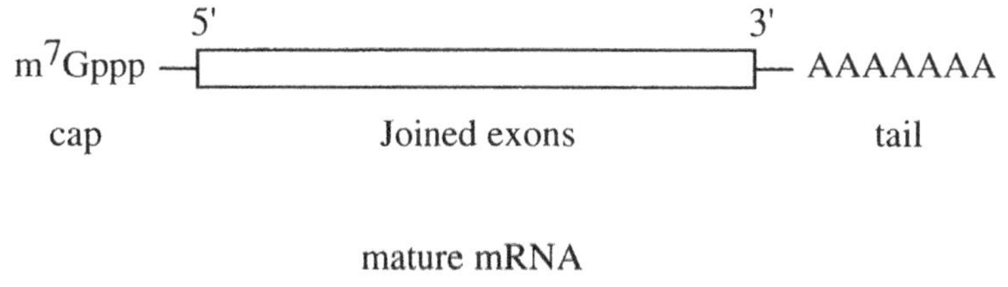

Chapter Test

True/False

1. The one gene–one polypeptide hypothesis states that for each polypeptide produced, a different set of mRNA, rRNA, and tRNA molecules must be synthesized.
2. One enzyme is responsible for copying DNA into RNA in bacteria; in eukaryotes, three enzymes are involved.
3. In eukaryotes, identical promoter sequences precede all genes. Because the promoter sequences are identical, accessory proteins are needed to ensure that only RNA polymerase II transcribes the genes that produce mRNAs.

4. DNA is double stranded. One strand is called the "coding" strand and the other the "noncoding" strand. The noncoding strand is used as the template to make mRNA. The relationship between the base sequence of the coding strand and the base sequence of the mRNA, ignoring the fact that the mRNA will contain uracil instead of thymine, is complementary.

Multiple Choice

5. Which of the following is not needed by bacterial RNA polymerase?
   a. A sigma subunit.
   b. A primer.
   c. GTP, CTP, UTP, and ATP.
   d. A promoter sequence.
   e. A DNA template.

6. Which one of the following statements about DNA and RNA synthesis is incorrect?
   a. Both DNA and RNA polymerases carry out synthesis of their respective nucleic acids in the 5′ to 3′ direction only.
   b. DNA polymerases use dNTPs to make DNA, whereas RNA polymerases use NTPs.
   c. The new DNA synthesized by DNA polymerases usually remains hydrogen bonded to the template strand, whereas the new RNA synthesized by RNA polymerases usually peels off the template strand.
   d. DNA polymerases require a single-stranded DNA template, whereas RNA polymerases require a single-stranded RNA template.
   e. Synthesis of both DNA and RNA is driven by energy generated by the release of the two terminal phosphate groups of the dNTP or NTP precursors.

7. The nucleotide sequence of one DNA strand of a double helix is as follows:

   5′-GCCTAGCAACAG-3′

   X = The 5′ to 3′ sequence of the complementary DNA strand;

   Y = The 5′ to 3′ sequence of the messenger RNA (mRNA) transcribed from this segment of DNA (i.e., Y = the sequence of the mRNA made from 5′-GCCTAGCAACAG-3′), where

   a. X = 5′-CGGATCGTTGTC-3′; Y = 5′-CUGUUGCUAGGC-3′
   b. X = 5′-CGGATCGTTGTC-3′; Y = 5′-CGGAUCGUUGUC-3′
   c. X = 5′-GACAACGATCCG-3′; Y = 5′-CUGUUGCUAGGC-3′
   d. X = 5′-CTGTTGCTAGGC-3′; Y = 5′-CUGUUGCUAGGC-3′
   e. X = 5′-CTGTTGCTAGGC-3′; Y = 5′-CTGTTGCTAGGC-3′

8. The RNA strand synthesized during transcription elongates until
   a. the entire chromosome has been copied into RNA.
   b. the RNA polymerase runs into the next gene.
   c. an intron is encountered on the DNA template strand.
   d. the RNA polymerase runs out of single-stranded DNA template.
   e. a specific termination sequence is reached on the DNA template strand.

Short Answer

9. Briefly describe the difference between bacterial and eukaryotic transcriptional termination.
10. Choose the item from each pair below that correctly describes transcription in eukaryotes.
    a. Single RNA polymerase, multiple RNA polymerases
    b. Noncoding sequences within and between genes, noncoding sequences only between genes
    c. Coupled transcription and translation, compartmentalized transcription and translation
11. If minimal media contains only a carbon source, salts, and some vitamins, why can wild-type *Neurospora* cultures such as those used in Beadle and Tatum's experiment survive on it?
12. Noncoding DNA found between genes is often referred to as ___________.

Essay

13. Are any regions of an intron's nucleotide sequence of importance? If so, what regions, and why are they important? Can you hypothesize what the effect might be if these regions were altered?

Chapter Test Answers

1. **False**
2. **True**
3. **False**
4. **False**
5. **b** 6. **d** 7. **d** 8. **e**
9. In bacteria, transcription termination occurs in one of two ways. In intrinsic or rho-dependent termination, the RNA polymerase recognizes a specific termination sequence in the DNA that signals the end of transcription. In rho-dependent termination, a special termination protein recognizes the termination sequence and causes the RNA polymerase to be released from the DNA template.
10. **a.** Multiple RNA polymerases; **b.** noncoding sequences within and between genes; **c.** compartmentalized transcription and translation.
11. Wild-type *Neurospora* cultures can grow on minimal media because they can produce all their own amino acids and can therefore make whatever proteins they require for their own metabolism.
12. Spacer DNA
13. Yes, short sequences near each end of the intron are important. These sequences are important in nuclear primary mRNA transcripts because they instruct the snRNPs where to cut the DNA to excise the introns. This excision must be precise for the intron to be removed without affecting the neighboring exons. If one of these sequences was altered, a

possible effect would be an imprecise splicing of the intron or no removal of the intron at all. This would in turn cause the translational apparatus to assume that the intron was intended to be part of the protein, leading to a larger and (most likely) nonfunctional protein. We now know that some internal intron sequences are translated. However, intron sequences that are translated are very rare and are not covered in this review.

Check Your Performance

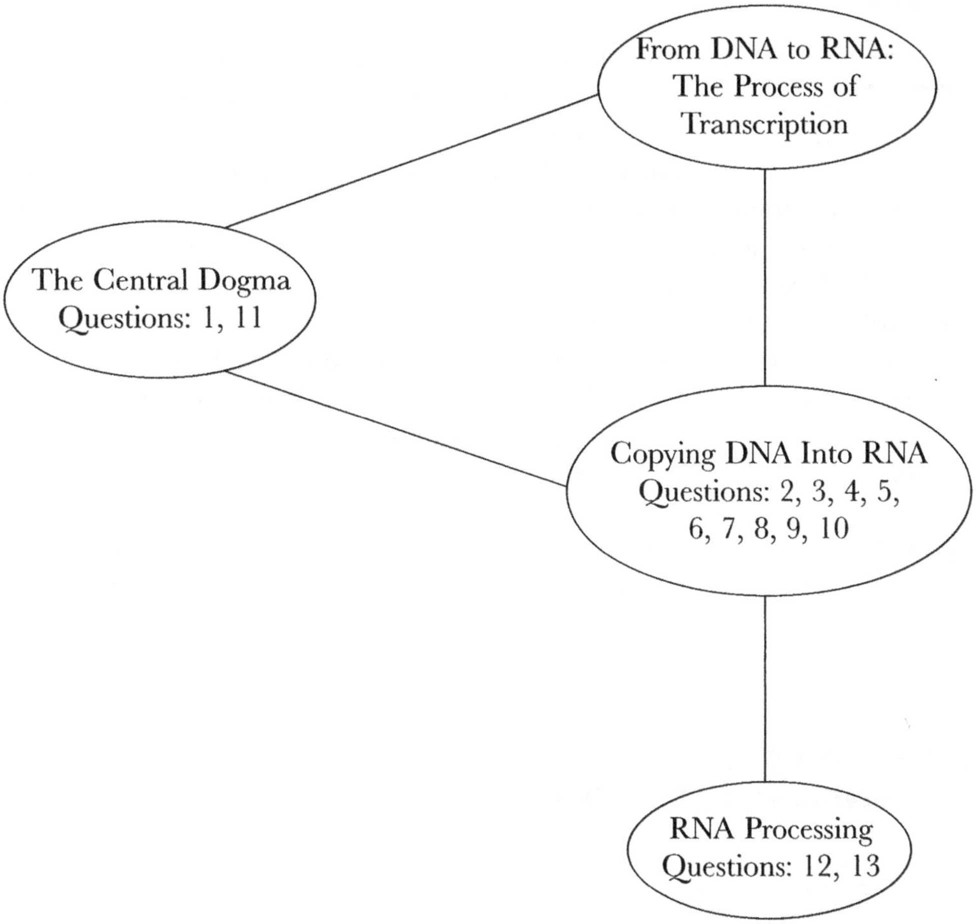

Note the number of questions in each grouping that you got wrong on the chapter test. Identify where you need further review and go back to relevant parts of this chapter.

CHAPTER 3

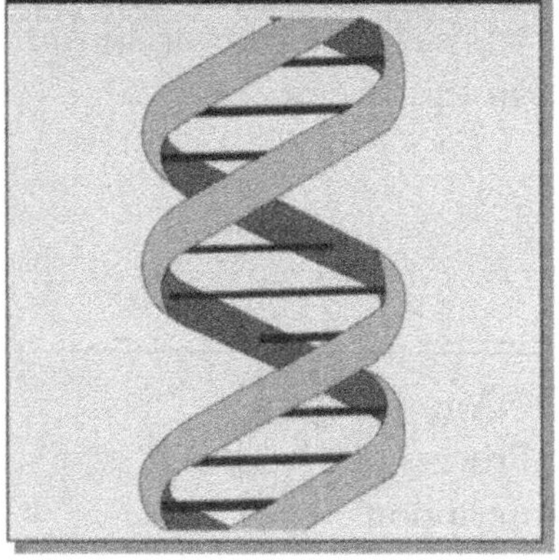

From Messenger RNA to Protein: The Process of Translation

Only four different nucleotides are found in messenger RNA (mRNA), but 20 different amino acids are commonly found in proteins. Therefore, it was immediately obvious that there cannot be a one-to-one correspondence between a nucleotide in mRNA and an amino acid in protein. The minimum number of nucleotides capable of specifying 20 different amino acids is three ($4^3 = 64$ possible different combinations), and a number of elegant experiments by M. W. Nirenberg, J. H. Matthaei, H. G. Khorana, and others showed that, indeed, the genetic code is a **triplet code**: three nucleotides specifying one amino acid. To determine the relationship between the nucleotide sequence in mRNA and the amino acid composition of a protein, Nirenberg synthesized an artificial mRNA consisting entirely of uracils and then demonstrated that this poly-U mRNA was translated into a polypeptide containing only one kind of amino acid: phenylalanine (Phe). The number of Phe residues obtained suggested that three uracils were needed to specify one Phe residue; in other words, the genetic code is a triplet code. A group of three nucleotides that specifies an amino acid is known as a **codon**, and 61 different codons specify amino acids; the remaining three codons signal termination of protein synthesis.

ESSENTIAL BACKGROUND

- Structure of proteins
- Structure of RNA
- Transcription

TOPIC 1: THE GENETIC CODE

KEY POINTS

✓ *How can the information carried in the* nucleotide sequence *of DNA specify the* amino acid *composition of proteins?*

✓ *There are threefold more codons than there are different amino acids found in cellular proteins. What are these "extra" codons used for?*

✓ *What does the universality of the genetic code suggest?*

THE GENETIC CODE IS DEGENERATE

There are 61 different codons that specify amino acids, but only 20 different amino acids are commonly found in cellular proteins. What are the remaining 41 codons used for? In fact, only two amino acids are specified by a single codon; all other amino acids are specified by multiple codons. For example, the amino acid serine (Ser) is specified by six different codons, the amino acid proline (Pro) by 4, and so on (**Table 3.1**). Because the genetic code allows different codons to specify the same amino acid, it is said to be a **degenerate** code.

THE GENETIC CODE IS UNIVERSAL

The same genetic code appears to be used by virtually all present-day organisms, from the simplest bacteria to the most complex plants and animals. A few codon differences have been observed in the DNA of some bacteria and protists and in the DNA of mitochondria and chloroplasts of higher organisms, but these differences are few and minor. The universality of the genetic code suggests that this common genetic language must have evolved very early in the history of life, early enough for it to be present in the common ancestors of present-day organisms.

Topic Test 1: The Genetic Code

True/False

1. A protein composed of 300 amino acids would be encoded by an mRNA of 100 nucleotides.
2. The same genetic code is used by virtually all organisms.
3. A codon with an A in the first position, a U in the second position, and any base in the third position will always code for isoleucine (Ile).

Table 3.1 The Genetic Code Dictionary

	U	C	A	G	
U	Phe	Ser	Tyr	Cys	**U**
	Phe	Ser	Tyr	Cys	**C**
	Leu	Ser	Stop	Stop	**A**
	Leu	Ser	Stop	Trp	**G**
C	Leu	Pro	His	Arg	**U**
	Leu	Pro	His	Arg	**C**
	Leu	Pro	Gln	Arg	**A**
	Leu	Pro	Gln	Arg	**G**
A	Ile	Thr	Asn	Ser	**U**
	Ile	Thr	Asn	Ser	**C**
	Ile	Thr	Lys	Arg	**A**
	Met	Thr	Lys	Arg	**G**
G	Val	Ala	Asp	Gly	**U**
	Val	Ala	Asp	Gly	**C**
	Val	Ala	Glu	Gly	**A**
	Val	Ala	Glu	Gly	**G**

Multiple Choice

4. A new organism is discovered. The structure of its DNA is similar to other organisms except that it contains six bases instead of the usual four. Furthermore, the proteins in the new organism contain 32 amino acids instead of 20. What is the minimum number of bases that could code for one amino acid in the new organism?
   a. 1
   b. 2
   c. 3
   d. 4
   e. 6

5. The genetic code is said to be **degenerate** because
   a. more than one triplet of nucleotides can code for a single amino acid.
   b. the same triplet of nucleotides can code for different amino acids.
   c. once mRNA is translated, it is degraded.
   d. there are more different amino acids than there are different codons.
   e. there are many steps between the synthesis of DNA and synthesis of a functional protein.

Short Answer

6. A bacterial gene 600 nucleotides long (not counting the stop codon) can code for a polypeptide chain of about how many amino acids and why?

7. Nirenberg synthesized an artificial mRNA consisting entirely of uracil residues (poly-U) and found that it resulted in a polypeptide composed of only the amino acid Phe. If he had synthesized an mRNA consisting of alternating uracil and cytosine residues (poly-UC), what amino acid(s) would the resulting polypeptide be composed of?

Topic Test 1: Answers

1. **False.** There are three nucleotides in a single codon, and each codon specifies a single amino acid, so a protein of 300 amino acids would be encoded by an mRNA of 900 nucleotides.

2. **True.** This is known as the "universality" of the genetic code. Any deviations from the genetic code are minor and rare.

3. **False.** A codon with an A in the first position, a U in the second position, and either an A, C, or U in the third position will code for isoleucine. However, if the third codon position is occupied by a G, the resulting AUG codon specifies methionine. AUG is in fact the only codon specifying methionine.

4. **b.** It was postulated that our genetic code is a triplet code (three bases specifying one amino acid) as follows: 20 amino acids are commonly found in proteins, so a one base code is not sufficient because it would provide for only four possible amino acids ($4^1 = 4$). A two base code is not enough either, as $4^2 = 16$, which is still not enough to account for the 20 known amino acids. Only a minimum of a three base code provides enough possibilities to account for 20 different amino acids ($4^3 = 64$). We use the same rationale to determine the genetic code for the new organism. A one base code would not be sufficient

because it would provide for only six possible amino acids ($6^1 = 6$). However, a two base code would be sufficient, yielding $6^2 = 36$ possibilities, which is more than the number of amino acids found in the organism.

5. **a.** There are 61 different codons that specify amino acids, but only 20 amino acids so d is incorrect. Many codons specify the same amino acid, but the opposite is never true—no individual codon specifies more than one amino acid, so b is incorrect. e is not a false statement, it is simply an incorrect answer for this question because it does not address the issue of degeneracy. c, although also not addressing the actual question, is only partially true, because mRNA molecules can be translated by many ribosomes before eventually being degraded.
6. 200. Each amino acid is specified by one codon, which consists of three nucleotides. A 600 nucleotide gene would therefore contain 600/3 = 200 codons and specify 200 amino acids.
7. A single polypeptide composed of serine-leucine-serine-leucine-etc. An mRNA consisting of alternating U and C residues carries only two codons, UCU and CUC. UCU specifies serine, and CUC specifies leucine.

TOPIC 2: COMPONENTS OF PROTEIN SYNTHESIS

KEY POINTS

✓ *Why can't a ribosome and an RNA polymerase molecule be present on a eukaryotic mRNA at the same time?*

✓ *Why can transcription and translation occur simultaneously in bacteria but not in eukaryotes?*

✓ *Does the cell need a different transfer RNA molecule for each of the 61 codons that specify amino acids?*

✓ *Can a single transfer RNA (tRNA) molecule that carries the amino acid serine base pair to all six codons that specify serine?*

mRNA CARRIES THE INFORMATION THAT DIRECTS PROTEIN SYNTHESIS

In translation, the nucleotide sequence information of an mRNA is translated into a sequence of amino acids in a protein. Eukaryotic mRNAs are synthesized in the nucleus but are translated into protein in the cytoplasm. Therefore, eukaryotic mRNAs have to move from the nucleus into the cytoplasm, a journey that cannot take place until transcription and processing of the eukaryotic mRNA are complete. Because bacteria do not have a nucleus, bacterial mRNAs, on the other hand, are synthesized and translated in the cytoplasm and transcription and translation of a bacterial mRNA can sometimes occur simultaneously.

PROTEIN SYNTHESIS OCCURS ON THE SURFACE OF RIBOSOMES

Ribosomes are complexes of ribosomal RNA (rRNA) and proteins located in the cytoplasm of eukaryotic and bacterial cells. They consist of a small subunit that binds an mRNA molecule and a large subunit that joins the mRNA/small subunit complex to form a complete ribosome. The large ribosomal subunit has a so-called P site (short for "peptidyl" site) that holds the growing polypeptide chain and an "A site" (short for "aminoacyl" site) that binds the next amino acid to be added to the growing polypeptide chain. The addition of the new amino acid in the A site to the growing polypeptide chain is catalyzed by one of the enzymes that form the large ribosomal subunit. Interestingly, many antibiotics work by blocking various steps in protein synthesis by binding to specific sites on eukaryotic or bacterial ribosomes (see the In the Clinic, below).

tRNA MOLECULES CARRY AMINO ACIDS TO mRNA ON THE RIBOSOME

The amino acids used in protein synthesis never directly contact the mRNA template that is attached to the small ribosomal subunit. Instead, the amino acids are aligned on the mRNA by means of tRNA molecules. These short single-stranded RNAs fold into L-shaped molecules by hydrogen bonding between complementary base sequences. An imaginary two-dimensional representation of a tRNA molecule is shown in **Figure 3.1**.

The middle loop (located at the "bottom" of the tRNA molecule) is the site of attachment to the mRNA template; the 3′ end of the tRNA (at the "top" of the molecule) is the attachment site for its amino acid.

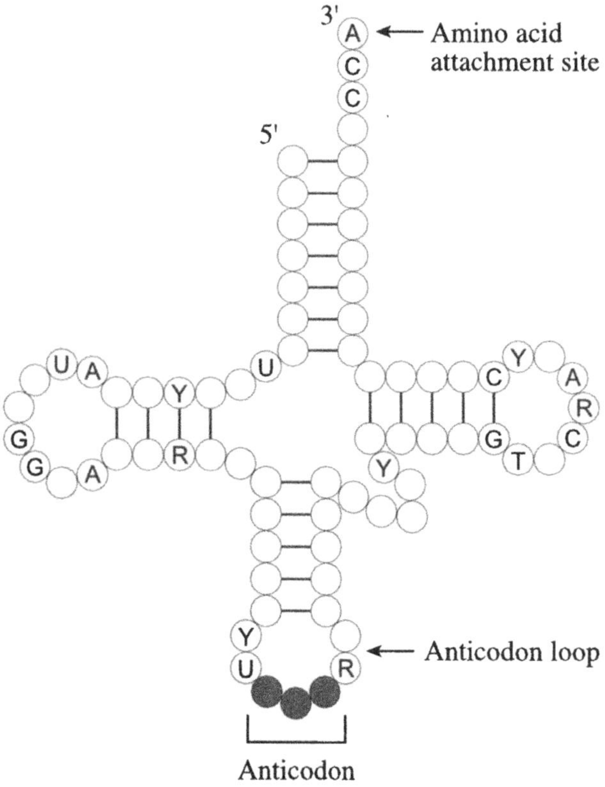

Figure 3.1 Structure of a tRNA molecule. A two-dimensional representation of a typical tRNA molecule is shown, with the anticodon loop positioned at the bottom. The anticodon is identified by the gray circles. Circles with letters are bases that are invariant in sequence among different tRNAs. A, C, G, U, and T indicate adenine, cytosine, guanine, uracil, and thymine, respectively. Y indicates that a pyrimidine is always found at that location; R indicates the position of a purine. The attachment site for amino acids is labeled, at the 3′ end of the molecule.

A tRNA MOLECULE ATTACHES TO THE mRNA TEMPLATE BY MEANS OF ITS ANTICODON

tRNA molecules carrying the amino acid serine base pair to mRNA codons that specify serine, tRNAs carrying the amino acid leucine base pair to codons that specify leucine, and so on. The part of the tRNA molecule that is responsible for this specific attachment to the mRNA template is known as the **anticodon**, a specific sequence of three bases located at the bottom of the tRNA molecule in the middle loop (Figure 3.1). Complementary base pairing between the anticodon (tRNA) and codon (mRNA) ensures that the correct amino acid is aligned on the mRNA template, although, as explained below, the formation of non–Watson-Crick base pairs plays an important role in protein synthesis.

SOME tRNA MOLECULES CAN BASE PAIR TO MORE THAN ONE CODON

Sixty-one of the 64 codons specify amino acids, but many organisms studied contain less than 61 different tRNA molecules (only 46 tRNA molecules with different anticodons have been identified in yeast). How are amino acids added to the codons for which there are no tRNA molecules? tRNA molecules are unusual in that some tRNA anticodons can form non–Watson-Crick base pairs (in other words, base pairs other than AT and GC base pairs). For example, despite the fact that there are two different codons that specify the amino acid cysteine (UGU and UGC), only one tRNA molecule is needed to carry cysteine because the anticodon of this tRNA, ACG, is able to base pair to *both* codons that specify cysteine. The ability of tRNA molecules to form non–Watson-Crick base pairs with the third base of the mRNA codon is referred to as **wobble**. Specific "wobble" rules ensure that a single tRNA molecule can only base pair to different codons that specify the *same* amino acid. Because most codons that specify the same amino acid differ only in the third base, this explains why wobble only occurs at the third position of the mRNA codon (**Figure 3.2**). A modified base called inosine (I), which is often found in the wobble position of tRNA anticodons, allows the maximum amount of wobble because it can base pair with cytosine (C), adenine (A), or uracil (U).

AMINOACYL-tRNA SYNTHETASES CATALYZE BOND FORMATION BETWEEN tRNA MOLECULES AND SPECIFIC AMINO ACIDS

Each amino acid has a specific aminoacyl-tRNA synthetase enzyme that catalyzes its attachment to a specific tRNA molecule. The formation of a tRNA–amino acid complex is an energy-requiring process driven by the hydrolysis of ATP. As a result, the bond that is formed between the tRNA and its amino acid is a high-energy bond and a tRNA molecule carrying an amino acid is said to be "**charged**." The energy trapped in the bond between the tRNA and its amino acid is later used to drive formation of the peptide bond that will join that amino acid to the growing protein chain.

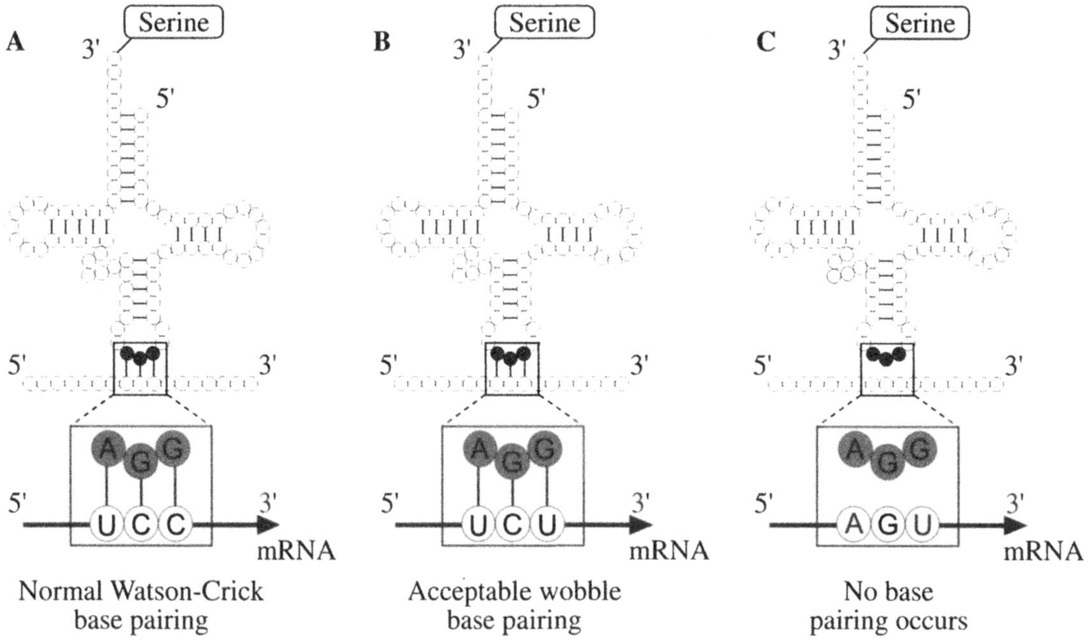

Figure 3.2 Wobble base pairing. In A, B, and C, the same tRNA molecule charged with the amino acid serine approaches three different mRNA codons specifying serine (UCC, UCU, and AGU all specify serine). In A, the 3′-AGG-5′ anticodon forms normal Watson-Crick base pairs with the 5′-UCC-3′ mRNA codon. In B, the anticodon encounters an mRNA codon that differs from the codon in A in the third position only. Because base pairing in the third position of the mRNA codon is relaxed, an unusual GU base pair is allowed. This phenomenon is referred to as **wobble**. However, when the anticodon approaches a codon that differs in the first or second positions of the mRNA codon (C), the formation of non–Watson-Crick base pairs is not allowed at these positions so a different tRNA molecule is needed to bring a serine to the AGU codon.

Topic Test 2: Components of Protein Synthesis

True/False

1. Many antibiotics used to treat bacterial infections work by selectively inhibiting bacterial protein synthesis.
2. It is possible for an RNA polymerase molecule and a ribosome to be attached to a eukaryotic mRNA simultaneously.
3. As opposed to mRNA and tRNA molecules that are composed only of the standard RNA bases, rRNA molecules often contain bases other than A, G, C, and U.

Multiple Choice

4. The three types of RNA responsible for forming part of the structure upon which protein synthesis occurs, dictating the order of amino acids in the growing protein chain, and transporting amino acids to the site of protein synthesis, respectively, are
   a. mRNA, rRNA, tRNA
   b. rRNA, tRNA, mRNA
   c. mRNA, tRNA, rRNA
   d. tRNA, rRNA, mRNA
   e. rRNA, mRNA, tRNA

Short Answer

5. The ________ in a tRNA molecule base pairs with a group of three nucleotides called the ________ in the mRNA template molecule.
6. What is the role of aminoacyl-tRNA synthetases in translation?
7. If "wobble" allows a single tRNA carrying a specific amino acid to base pair with more than one codon specifying that amino acid, why is more than one tRNA molecule needed for some amino acids?

Topic Test 2: Answers

1. **True.** Structural and functional differences between bacterial and eukaryotic ribosomes allow antibiotics to specifically target bacterial ribosomes. The host eukaryotic (human) ribosomes are not affected.
2. **False.** RNA polymerase makes mRNA from a DNA template (in the process known as transcription) in the nucleus of a eukaryotic cell. Ribosomes carry out translation in the cytoplasm.
3. **False.** It is actually tRNA that can contain modified nucleotides. One of these modified nucleotides, called **inosine**, is often found in the anticodon position that base pairs with the third codon position of the mRNA.
4. **e.** rRNA molecules complex with ribosomal proteins to form ribosomes, the sites of protein synthesis. mRNA molecules carry the codon sequences that specify the amino acid composition of proteins. tRNA molecules are coupled to amino acids and act as "bridges" between the codon sequence of the mRNA and the amino acids that the codons specify.
5. anticodon; codon. Base pairing between the 5′ to 3′ sequence of the mRNA codon and the 3′ to 5′ sequence of the tRNA anticodon holds the tRNA on the ribosome long enough for the amino acid with which it is "charged" to be added to the growing protein chain.
6. Aminoacyl-tRNA synthetases are the enzymes that catalyze bond formation between tRNA molecules and specific amino acids in a process known as "charging."
7. Wobble only allows non–Watson-Crick base pairs to form at the third codon position in the mRNA. However, some codons that specify the same amino acid differ in the first position and/or second position. In these cases, wobbling is not permitted, and different tRNAs are required to translate these codons.

TOPIC 3: TRANSLATION: BUILDING A PROTEIN

Key Points

✓ *Does protein synthesis start at the very beginning of the mRNA with the very first codon?*

✓ *How does the ribosome know where on the mRNA to start synthesizing a protein?*

✓ *How many different types of protein can be made from a single mRNA: in bacteria? in eukaryotes?*

✓ *How are amino acids aligned on the mRNA template?*

There are three stages of protein synthesis: initiation, elongation, and termination.

Initiation

During the initiation stage, a small ribosomal subunit binds an initiator tRNA and an mRNA. The initiator tRNA carries the amino acid methionine (in eukaryotes) or a derivative of methionine called *N*-formyl-methionine (in bacteria). Thus, methionine, or a derivative of methionine, starts all protein chains (*N*-formyl-methionine is only used to *start* protein chains) and the initiator tRNA base pairs to the **initiator codon** AUG on the mRNA. With the help of proteins called initiation factors and the hydrolysis of one GTP molecule, the large ribosomal subunit then joins to the small subunit to form a complete **initiation complex**.

The codon AUG serves two roles: it directs the addition of the amino acid methionine during protein synthesis and it is also the codon that *starts* protein synthesis (the so-called initiator codon). The small ribosomal subunit has to distinguish between the two types of AUG codon. The mechanism used to locate initiator codons differs between bacteria and eukaryotes.

In most bacterial mRNAs, the initiator codon is preceded by a special sequence called a **ribosome binding site (RBS)** or **Shine-Dalgarno sequence**. The RBS sequence is complementary to an rRNA in the small ribosomal subunit. As the small subunit passes over the RBS, the complementary rRNA base pairs to it, positioning the initiator AUG so that it lies directly below the initiator tRNA in the P site of the large ribosomal subunit. AUGs that specify internal methionines are not preceded by RBS sequences.

A single bacterial mRNA can contain several RBS sequences because a bacterial mRNA can carry the information for more than one type of protein. An mRNA that gives rise to multiple different proteins is said to be **polycistronic**. Polycistronic mRNAs are found in bacteria for reasons that are explained in Chapter 5. (Eukaryotic nuclear DNAs do not give rise to polycistronic mRNAs. However, mRNAs that carry the information for more than one type of protein are found in the chloroplasts and mitochondria of eukaryotic cells. That chloroplast and mitochondrial DNAs give rise to polycistronic mRNAs supports the theory that chloroplasts and mitochondria originated from bacterial cells that were engulfed by the ancestors of today's eukaryotic cells and sequestered as internal organelles. See In the Clinic, below.)

In eukaryotes there is only one initiator AUG codon per nuclear mRNA, and hence only one type of protein is produced. Eukaryotic nuclear mRNAs are therefore **monocistronic**—each mRNA gives rise to a single protein type. To locate the initiator AUG, the small ribosomal subunit has to first locate the **5′-methyl-guanosine cap**, a bulky structure that gets added to the 5′ end of all eukaryotic mRNAs. The small subunit then slides down the mRNA until it locates an AUG codon. This AUG is used as the initiator codon (**Table 3.2**).

Table 3.2 Differences Between Bacterial and Eukaryotic Protein Synthesis

BACTERIA	EUKARYOTES
70S ribosomes	80S ribosomes
First amino acid = *N*-formyl methionine	First amino acid = methionine
Ribosome binding site (RBS) sequences	5′-Methylguanosine cap
Polycistronic mRNAs	Monocistronic mRNAs
Coupled transcription/translation	Compartmentalized transcription/translation
mRNA half-life = 1–3 min	mRNA half-life = 30 min–10 hr

The way in which the nucleotide sequence of an mRNA is divided into groups of three bases (codons) is unambiguously defined once the initiator codon is selected. Thus, during formation of the initiation complex, the nucleotide sequence that is actually translated into a protein, commonly referred to as the **reading frame**, is defined.

Elongation

In the initiation complex, the initiator tRNA occupies the P site of the large ribosomal subunit and the codon that specifies the second amino acid lies immediately below the A site. With the aid of a protein called **elongation factor-Tu** (or EF-Tu) and the energy released by the hydrolysis of one GTP molecule, the tRNA molecule carrying the second amino acid enters the A site of the ribosome and base pairs to the second mRNA codon. In the next step, the methionine (or *N*-formyl-methionine) leaves the initiator tRNA in the P site and forms a peptide bond with the second amino acid in the A site. This process is catalyzed by **peptidyl transferase**, which is a component of the large ribosomal subunit. Hydrolysis of the bond between the methionine and the initiator tRNA provides the energy that drives peptide bond formation. In the final step, the growing peptide chain that is now attached to the tRNA in the A site is "**translocated**" to the P site by the movement of the ribosome a distance of three bases toward the 3′ end of the mRNA. During this step, the now empty initiator tRNA is ejected from the ribosome, the growing peptide chain moves into the P site, and the codon specifying the third amino acid moves into the A site. Translocation requires energy from the hydrolysis of another GTP molecule and the presence of elongation factor EF-G. The elongation process then repeats itself: A tRNA molecule carrying the third amino acid base pairs to the mRNA codon in the A site; peptidyl transferase catalyzes peptide bond formation between the dipeptide in the P site and the amino acid in the A site, leaving an empty tRNA molecule in the P site; the ribosome translocates along the mRNA, ejecting the empty tRNA, returning the growing peptide chain (now a tripeptide) to the P site and bringing a new mRNA codon into the A site in preparation for the addition of the fourth amino acid (**Figure 3.3**).

Termination

Termination of protein synthesis occurs when one of the three stop codons—UAA, UGA, or UAG—enters the A site of the ribosome. There are no tRNA molecules that recognize the stop codons. Instead, a protein called a **release factor** binds to the A site and causes hydrolysis of the bond holding the completed polypeptide chain to the tRNA in the P site. The completed polypeptide is released and the ribosome separates into its small and large subunits in preparation for the next round of protein synthesis.

PROTEINS ARE SYNTHESIZED FROM THE N TERMINUS TO THE C TERMINUS

Every amino acid contains an amino (—NH_2) group and a carboxyl (—COOH) group. When two amino acids join together to form a peptide bond, the amino group of one amino acid joins to the carboxyl group of the other amino acid. The resulting dipeptide still has an NH_2 group (denoted by the letter "N") at one end and a COOH group (denoted by the letter "C") at the other end.

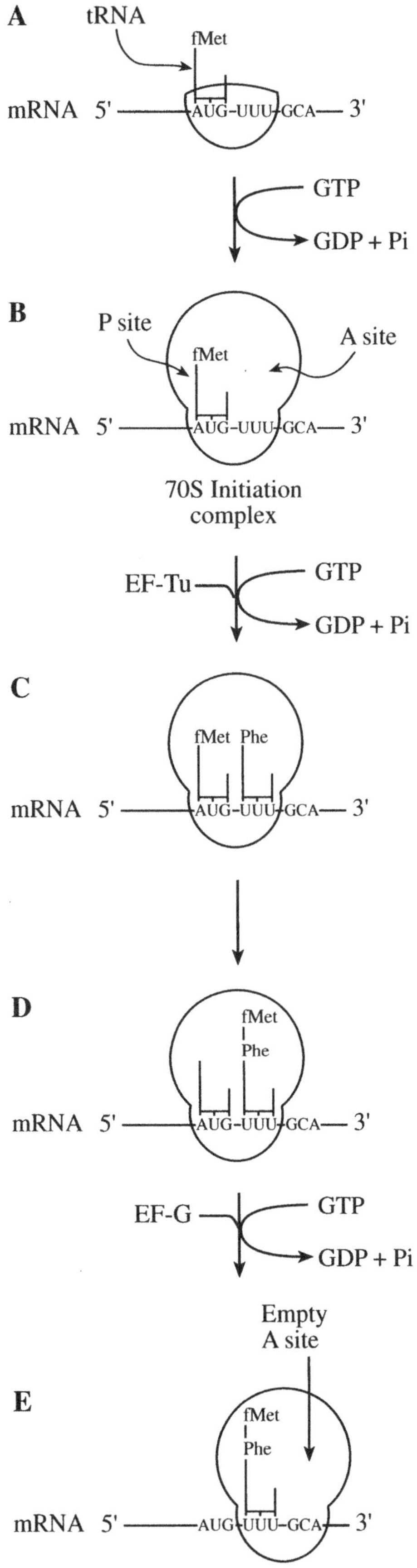

Figure 3.3 Steps in bacterial protein synthesis. (A) An mRNA, the small ribosomal subunit, and a tRNA charged with *N*-formyl methionine join together in the presence of initiation factors and GTP. (B) Hydrolysis of GTP provides the energy required to add the large ribosomal subunit. The initiation factors are released and the completed complex is call the 70S Initiation complex. The tRNA molecule charged with *N*-formyl methionine resides in the P site, whereas the A site is empty. (C) GTP hydrolysis again provides energy, this time for binding of a new charged tRNA to the empty A site. (D) The amino acid in the P site leaves its tRNA and covalently attaches via a peptide bond to the amino acid attached to the tRNA in the A site. This reaction is catalyzed by the enzyme peptidyl transferase using energy released by breakage of the bond between the *N*-formyl methionine and its tRNA. (E) "Translocation" of the ribosome toward the 3′ end of the mRNA. Energy is again provided by hydrolysis of GTP. The ribosome shifts three bases down the mRNA toward the 3′ end. The uncharged tRNA in the P site is released, the tRNA with the growing polypeptide chain moves into the P site, and a new codon enters the A site ready for the binding of a new charged tRNA.

The next amino acid would add on to the carboxyl group of this dipeptide. Thus, proteins are said to be synthesized in the N to C direction to signify the fact that new amino acids always add on to the carboxyl or C terminal end of the growing protein chain.

MANY COPIES OF A PROTEIN ARE USUALLY MADE

Most proteins are synthesized in many copies. To facilitate the production of multiple copies of a protein, a single mRNA can be translated simultaneously by several ribosomes, creating the appearance in an electron microscope of beads (the ribosomes) on a string (the mRNA). A cluster of ribosomes simultaneously translating a single mRNA is referred to as a **polysome** (**Figure 3.4**).

Topic Test 3: Translation: Building a Protein

True/False

1. If the anticodon of a $tRNA^{His}$ was modified by a single-base change to recognize an Arg codon and then added to a translation system in the presence of $tRNA^{Arg}$, the resulting protein would have histidine at all positions normally occupied by arginine.
2. The amino acid that starts protein chains in eukaryotic protein synthesis is found only at the N-terminus of proteins.
3. At no time during protein synthesis does an amino acid make direct contact with the mRNA template.

Multiple Choice

4. Which one of the following statements is correct?
   a. A single eukaryotic nuclear mRNA molecule can carry the information for several different proteins.
   b. Polysomes produce many copies of the same protein from a single mRNA molecule.

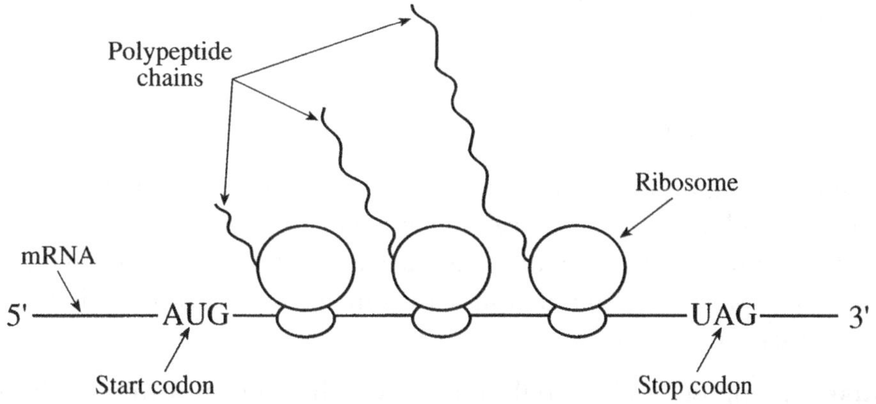

Figure 3.4 Rapid synthesis of multiple copies of a protein. An mRNA is shown with a start codon and a stop codon. Multiple ribosomes are capable of synthesizing copies of the protein simultaneously as shown in the figure, a structure known as a **polysome**.

c. When one of the stop codons enters the A site on the ribosome, a special tRNA molecule causes termination of protein synthesis.
d. There have to be 64 different tRNA molecules because 64 different codons are specified by the genetic code.
e. All of the above

5. Which of the following statements about protein synthesis is correct?
   a. During protein chain elongation, the ribosome moves closer to the 5′ end of the mRNA template.
   b. The growing protein chain is passed from the tRNA molecule in the P (peptidyl) site to the amino acid attached to the tRNA molecule in the A (aminoacyl) site.
   c. The tRNA molecule carrying the next amino acid to be added to the growing protein chain enters the empty P site.
   d. The tRNA molecule in the A site moves down three bases (one codon) toward the 3′ end of the mRNA.
   e. Proteins are synthesized from the carboxy (C)-terminus to the amino (N)-terminus.

6. Which of the following does not apply to bacterial mRNAs?
   a. Shine-Dalgarno or Ribosome binding site sequence
   b. Coupled transcription and translation
   c. Ability to carry the information for more than one gene on a single mRNA
   d. Stop codons
   e. 5′-Methyl guanosine cap

Short Answer

7. One of the frames in an animation of protein synthesis shows a ribosome holding a tRNA molecule. The tRNA has a tripeptide (a protein chain three amino acids long) attached to it and it is located in the P site of the ribosome—the A site is empty. What would the next frame in the animation show?

8. One of the frames in an animation of protein synthesis shows a ribosome holding *two* tRNA molecules. One tRNA has a tripeptide (a protein chain three amino acids long) attached to it and it is located in the A site of the ribosome. The other tRNA is *not* attached to an amino acid and it is located in the P site of the ribosome. What would the next frame in the animation show?

Topic Test 3: Answers

1. **False.** Although it is true that this modified $tRNA^{His}$ would recognize codons for arginine and put histidine in place of arginine, it is not the only tRNA molecule capable of recognizing the arginine codon. The question states that the translation system also includes $tRNA^{Arg}$, ensuring that at least some of the codons specifying arginine will result in the insertion of arginine.

2. **False.** As opposed to bacterial proteins, which all begin with a modified form of the amino acid methionine (*N*-formyl methionine), eukaryotic proteins begin with a standard methionine amino acid.

3. **True.** Amino acids are physically separated from the mRNA by the tRNA to which they are coupled. The tRNA anticodon that contacts the mRNA is at the "bottom" of the tRNA cloverleaf structure, whereas the amino acid attachment site is at the "top."
4. **b.** Eukaryotic nuclear mRNAs are monocistronic, meaning they carry the information for only one kind of protein, so a is incorrect. c is incorrect because stop codons are not recognized by tRNA molecules but rather by proteins called release factors. And d is not correct because there are usually fewer than 64 tRNA molecules due to "wobble" and the fact that 3 of the 64 codons are stop codons.
5. **b.** Ribosomes move from the 5′ end of the mRNA to the 3′ end and synthesize proteins from the amino to the carboxy terminus; this eliminates a and e. c is not correct because the new tRNA carrying the next amino acid enters the A site of the ribosome, not the P site. d is incorrect because the tRNA molecule does not move once it enters the A site, until it is ejected from the ribosome. The tRNA anticodon pairs with the mRNA codon and remains that way, whereas the ribosome shifts position, moving the tRNA into the P site.
6. **e.** In bacterial mRNAs, the small ribosomal subunit uses ribosome binding site sequences to locate initiator AUGs. The 5′-methylguanosine cap structure is used to locate the initiator AUG in eukaryotic mRNAs. All other options are true for bacterial mRNAs.
7. The next frame would show a new tRNA charged with a single amino acid entering the A site of the ribosome.
8. The next frame would show the tRNA in the P site leaving the ribosome, as the ribosome moves toward the 3′ end of the mRNA, shifting the tRNA with the tripeptide into the P site. The A site is now open and ready for a new tRNA carrying the fourth amino acid.

IN THE CLINIC

A particular antibiotic has clinical value only if it acts on bacteria and not on animal cells. Many antibiotics used to ward off bacterial infections work by blocking various steps in protein synthesis. Functionally, bacterial and eukaryotic ribosomes are very similar, but their structures differ sufficiently that antibiotics can recognize and bind specifically to bacterial ribosomes. For example, streptomycin binds to a specific protein in the small ribosomal subunit and prevents binding of the initiator tRNA, whereas tetracycline and kanamycin, which also bind to the small ribosomal subunit, inhibit binding of a charged tRNA to the A site and translocation, respectively. Chloramphenicol binds to the large ribosomal subunit and inhibits peptidyl transferase. The equivalent steps in eukaryotic protein synthesis are not affected by these antibiotics.

Interestingly, mitochondrial ribosomes (mitochondria have their own DNA and can carry out their own protein synthesis) are sensitive to many of the same antibiotics used to inhibit protein synthesis in bacteria. This was an especially important observation because it helped to support the theory that mitochondria (and chloroplasts) originated from bacterial cells that were engulfed by the ancestors of today's eukaryotic cells and sequestered as internal organelles.

DEMONSTRATION PROBLEM

One strand of a section of DNA isolated from *Pseudomonas syringae* reads 5′-GGCTATGT-TAAAAACTAA-3′.

Question

What would be the 5′ to 3′ nucleotide sequence of the mRNA transcribed from this DNA strand?

Solution

RNA polymerase makes RNA from a DNA template by adding complementary RNA bases opposite the DNA bases. The sequence of the mRNA is therefore the complement of the DNA, remembering that **uracil** is found in mRNA instead of **thymine**. However, note that the template sequence is written from 5′ to 3′. Because the mRNA strand will be antiparallel (run in the opposite direction) to the DNA strand, the sequence of the mRNA has to be reversed to give the correct answer to the question, which asks for the 5′ to 3′ sequence of the mRNA. For example,

5′-GGCTATGTTAAAAACTAA-3′ (this is the DNA template)
3′-CCGAUACAAUUUUUGAUU-5′ (this is the 3′ to 5′ mRNA sequence)
5′-UUAGUUUUUAACAUAGCC-3′ (this is the 5′ to 3′ mRNA sequence)

Question

What would be the 5′ to 3′ nucleotide sequence of the mRNA if RNA polymerase transcribed the complement of the DNA strand shown above?

Solution

The same procedure as described above is used, but first the sequence of the new template DNA strand has to be figured out. The sequence of the new template will be complementary and antiparallel (3′ to 5′) to the sequence given in the question. As before, RNA polymerase will make a complementary RNA strand but this time, the template sequence is running 3′ to 5′ so the mRNA sequence will be in the 5′ to 3′ direction, which is what the question asks for. As you can see, the mRNA sequence is identical to the *original* DNA template, with the exception that it contains uracils in place of the thymines.

5′-GGCTATGTTAAAAACTAA-3′ (this is the original DNA sequence)
3′-CCGATACAATTTTTGATT-5′ (this is the 3′ to 5′ DNA template)
5′-GGCUAUCUUAAAAACUAA-3′ (this is the 5′ to 3′ mRNA sequence)

Question

If the strand that is complementary to the original DNA strand is used as a template (as in the previous example), is this section of DNA most likely from the beginning of the gene or the end of the gene, assuming that the above sequence is in the correct reading frame?

Solution

We determined in answering the last question that the 5′ to 3′ sequence of the mRNA transcribed from the **complement** of the original strand is as follows:

5′ - GGCUAUCUUAAAAACUAA - 3′ (5′ to 3′ mRNA sequence)

If we assume that this sequence is in the correct reading frame, then beginning with the first base, the reading frame would specify the following codons:

5′-GGC-UAU-CUU-AAA-AAC-UAA-3′

The question asks if we can determine whether this sequence is most likely from the beginning or end of the gene, which means that we need to look for either an AUG start codon (suggesting the beginning of the gene) or one of the three stop codons, UAA, UAG, UGA (suggesting the end of the gene). No AUG codon is found in this reading frame. However, the last codon in this sequence is UAA, which is one of the stop codons. This suggests that this section of DNA is from the end of the gene.

Question

If the strand that is complementary to the given DNA sequence is used as a template, use the codon table to determine the amino acids encoded by the resulting mRNA.

Solution

As explained in the previous section, the 5′ to 3′ mRNA sequence transcribed from the complement of the original strand is as follows:

5′-GGC-UAU-CUU-AAA-AAC-UAA-3′

Using the codon table, it is determined that these codons correspond to the following amino acids:

GGC = Gly
UAU = Tyr
CUU = Leu
AAA = Lys
AAC = Asn
UAA = Stop

So the amino acid sequence of the resulting polypeptide would be

N–Gly-Tyr-Leu-Lys-Asn–C

	U	C	A	G	
U	Phe Phe Leu Leu	Ser Ser Ser Ser	Tyr Tyr Stop Stop	Cys Cys Stop Trp	U C A G
C	Leu Leu Leu Leu	Pro Pro Pro Pro	His His Gln Gln	Arg Arg Arg Arg	U C A G
A	Ile Ile Ile Met	Thr Thr Thr Thr	Asn Asn Lys Lys	Ser Ser Arg Arg	U C A G
G	Val Val Val Val	Ala Ala Ala Ala	Asp Asp Glu Glu	Gly Gly Gly Gly	U C A G

Chapter Test

True/False

1. Termination of translation occurs only once on a eukaryotic mRNA but can occur many times on a bacterial mRNA.
2. RBS (or Shine-Dalgarno) sequences are highly conserved nucleotide sequences in bacterial mRNAs that determine the sites where RNA synthesis is initiated.
3. Each step involved in adding an amino acid to a growing protein chain requires energy.

Multiple Choice

4. Which one of the following statements about protein synthesis is incorrect?
   a. Eukaryotic nuclear mRNAs can carry the information for one type of protein only.
   b. With the exception of the initiator tRNA, tRNA molecules enter the A site on the ribosome and leave from the P site.
   c. In eukaryotes, a ribosome and an RNA polymerase molecule can be present on an mRNA at the same time.
   d. The energy that drives peptide bond formation during protein synthesis is obtained from the bond that attaches the amino acid to its tRNA molecule.
   e. Some tRNA molecules can base pair to two or even three different codons in the mRNA.
5. Choose the option below that correctly matches each of the terms numbered 1 through 5 with one of the descriptions listed a through e:
   1. Peptidyl transferase
   2. Wobble
   3. Aminoacyl-tRNA synthetases
   4. Translocation
   5. Release factor

a. Formation of non–Watson-Crick base pairs in the third position of codons
b. Movement of the ribosome on the mRNA template
c. Termination of protein synthesis
d. "Charging" of tRNA molecules
e. Peptide bond formation
  a. 1 = e, 2 = a, 3 = d, 4 = b, 5 = c
  b. 1 = e, 2 = b, 3 = d, 4 = a, 5 = c
  c. 1 = e, 2 = d, 3 = c, 4 = b, 5 = a
  d. 1 = d, 2 = a, 3 = e, 4 = b, 5 = c
  e. 1 = d, 2 = b, 3 = e, 4 = a, 5 = c

6. Which one of the following is not directly involved in the process known as translation?
   a. mRNA
   b. DNA
   c. Peptidyl transferase
   d. Ribosomes
   e. Aminoacy l-tRNA synthetases

Short Answer

7. The order in which the triplets of bases in an mRNA is read is most precisely called the _________.

8. If wobble allowed a tRNA molecule to base pair to all four bases in the third position of the mRNA codon, what would be the minimum number of tRNA molecules required to carry the amino acid arginine?

Essay

9. Briefly describe how the start of translation is "signaled" in bacteria and eukaryotes. Include in your discussion an explanation of the terms monocistronic and polycistronic.

10. Name and describe the three major steps in translation.

Chapter Test Answers

1. **True**

2. **False**

3. **True**

4. **c** 5. **a** 6. **b**

7. Reading frame

8. Two

9. In bacteria, the small ribosomal subunit searches for a ribosome binding site or Shine-Dalgarno sequence. This is a sequence of six to eight nucleotides that precedes most initiator AUG codons. Translation then begins at the initiator AUG, where an *N*-formyl methionine is used to start the protein chain. There can be multiple Shine-Dalgarno sequences on a single bacterial mRNA molecule. In other words, a single bacterial mRNA

can carry the information for the synthesis of different proteins. Such mRNAs are said to be polycistronic. In eukaryotes, the small ribosomal subunit recognizes the 7-methylguanosine cap at the 5′ end of the mRNA and protein synthesis begins at the first AUG that the small ribosomal subunit encounters after the cap. The AUG closest to the cap structure is the only AUG that is used to start a protein chain in eukaryotic mRNAs. This means that each eukaryotic mRNA encodes only one protein type and is said to be monocistronic.

10. **Initiation:** Binding of the initiator tRNA (carrying either methionine or *N*-formyl methionine), the mRNA template, proteins called initiation factors, and GTP to a small ribosomal subunit. Joining of the large ribosomal subunit forms a complete initiation complex in a reaction driven by the hydrolysis of GTP. **Elongation:** Binding of a "charged" tRNA to the empty A site on the ribosome. Movement of the amino acid chain from the tRNA in the P site to the amino acid attached to the tRNA in the A site. Peptide bond formation by peptidyl transferase. Translocation of ribosome three bases closer to the 3′ end of the mRNA, resulting in ejection of uncharged tRNA from P site and movement of tRNA with growing peptide chain from A site to P site. **Termination:** Entrance of stop codon into A site. Binding of release factors. Hydrolysis of bond between completed protein chain and tRNA. Dissociation of ribosomal subunits.

Check Your Performance

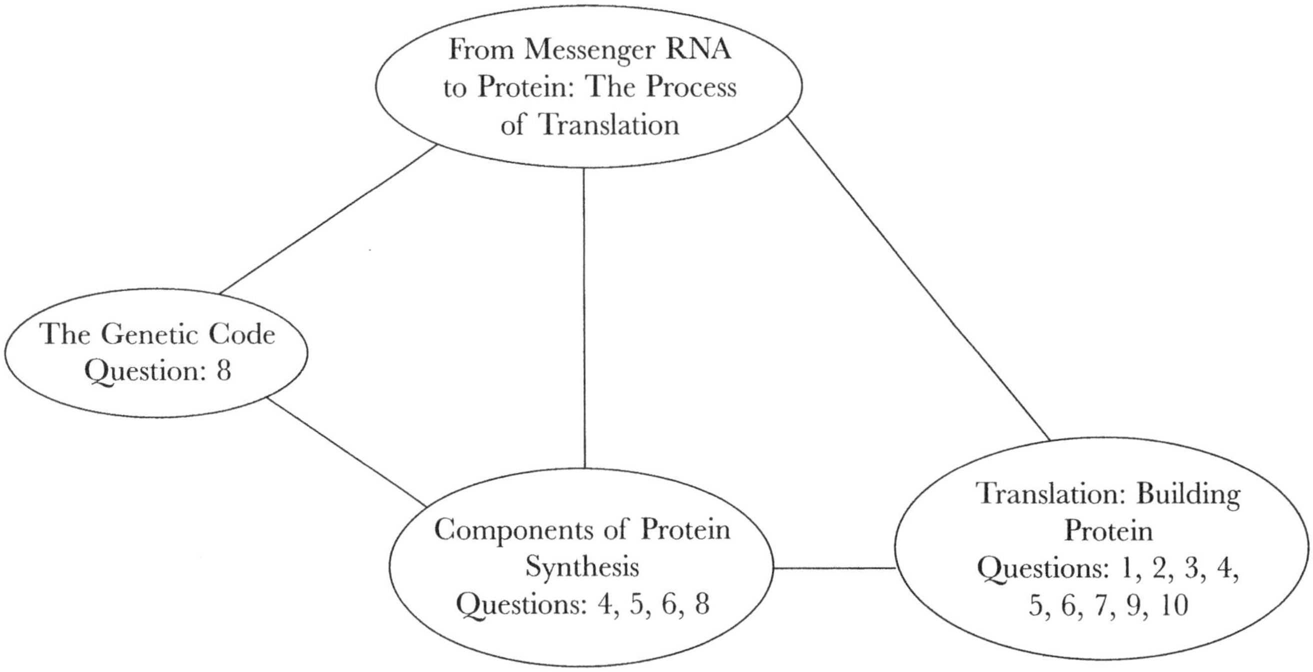

Note the number of questions in each grouping that you got wrong on the chapter test. Identify where you need further review and go back to relevant parts of this chapter.

CHAPTER 4

Mutations

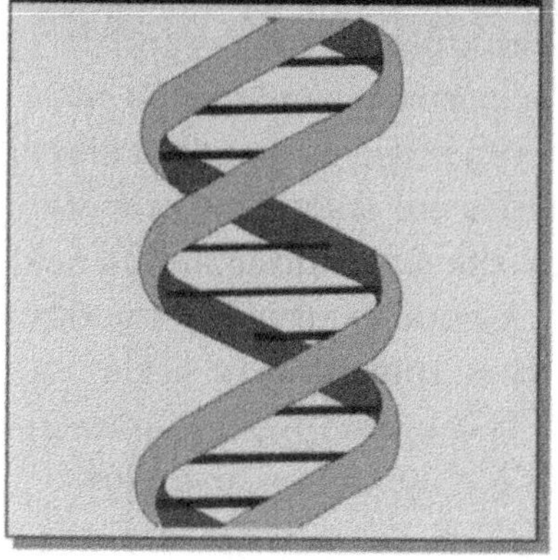

The **wild-type** form of a gene refers to the nucleotide sequence of that gene that is characteristic of most individuals of a species, also called the normal, standard, or reference genotype. **Mutations** are heritable changes in the genetic material caused by alterations in the nucleotide sequence in DNA. Such alterations give rise to alternate forms of genes, called **alleles**. Two broad types of mutations are recognized; **point mutations** and **chromosomal mutations**. Point mutations are mutations of single genes: One allele becomes another because of small alterations in the nucleotide sequence. Chromosomal mutations are those that affect whole chromosomes or segments of chromosomes, for example, by changing the position or direction of a segment of DNA.

ESSENTIAL BACKGROUND

- Structure of DNA
- DNA replication

TOPIC 1: TYPES OF GENE MUTATIONS

KEY POINTS

✓ *What is a mutation?*

✓ *What are the different types of mutations?*

✓ *What is a conditional mutant?*

✓ *Why do frameshift mutations almost always result in protein products that are shorter than the normal proteins?*

Point mutations are grouped into two broad types: **base substitution mutations** and **frameshift mutations**.

BASE SUBSTITUTION MUTATIONS CHANGE A SINGLE CODON

As their name implies, **base substitution mutations** are changes in the nucleotide sequence that substitute one base for another. They are categorized into three basic types, **missense**, **nonsense**, or **samesense**, based on the effect they have on the amino acid sequence in the resulting protein.

Missense mutations change the nucleotide sequence such that one amino acid is replaced by a different amino acid in the final protein (**Figure 4.1A**). The consequence of a missense mutation depends on whether the amino acid affected is essential for protein function and, if so, on the properties of the substituting amino acid. Thus, missense mutations affecting nonessential amino acids are often tolerated by the protein, which remains fully or partially functional. On the other hand, missense mutations that affect amino acids that are critical for protein function (such as amino acids in the active site of an enzyme or amino acids that are critical for proper folding of the protein) usually result in a nonfunctional protein unless the substituting amino acid has identical or very similar properties to the amino acid it replaces. A good example of the dire consequences of missense mutations is provided by the sickle cell anemia mutation. In sickle cell anemia, the sequence GAG in the gene for β-globin is changed to GTG. In the β-globin messenger RNA (mRNA), this mutation changes the codon GAG that codes for the amino acid glu-

WILD TYPE* CODING DNA AND mRNA SEQUENCES

ATG	CCG	TGT	CAG	ATG	TTC	– DNA
AUG	CCG	UGU	CAG	AUG	UUC	– mRNA
Met	Pro	Cys	Gln	Met	Phe	– Amino acid sequence

A. MISSENSE MUTATION - a codon is changed to a different codon that specifies a different amino acid

ATG	CCG	TGG	CAG	ATG	TTC	– DNA
AUG	CCG	UGG	CAG	AUG	UUC	– mRNA
Met	Pro	Trp	Gln	Met	Phe	– Amino acid sequence

B. SAMESENSE MUTATION - a codon is changed to a different codon that specifies the same amino acid

ATG	CCG	TGC	CAG	ATG	TTC	– DNA
AUG	CCG	UGC	CAG	AUG	UUC	– mRNA
Met	Pro	Cys	Gln	Met	Phe	– Amino acid sequence

C. NONSENSE MUTATION - a codon that specifies an amino acid is changed to a stop codon

ATG	CCG	TGA	CAG	ATG	TTC	– DNA
AUG	CCG	UGA	CAG	AUG	UUC	– mRNA
Met	Pro	STOP				– Amino acid sequence

* **Wild type is the genotype that is characteristic of the majority of individuals of a species, also called the normal, standard or reference genotype. In this example, the "wild type" sequence refers to the unmutated sequence.**

Figure 4.1 Base substitution mutations.

tamic acid to GUG that codes for valine. This single amino acid difference causes a profound structural distortion of hemoglobin, the oxygen-carrying protein in red blood cells that is composed of two polypeptide chains of β-globin and two of α-globin. Thus, normal hemoglobin remains soluble under normal physiological conditions, but the hemoglobin in sickle cells precipitates when the blood oxygen level falls, forming long fibrous aggregates that distort the blood cells into the sickle shape.

Finally, some missense mutations are harmful only under certain *restrictive* conditions and are not detectable under other *permissive* conditions. Organisms carrying such mutations are called **conditional mutants**. An example of conditional mutants are heat-sensitive mutants. Heat-sensitive *Escherichia coli* mutants are able to grow at a lower permissive temperature such as 30°C but are unable to grow at a higher restrictive temperature such as 37°C, probably due to an increased tendency of the mutant protein to denature at restrictive temperatures.

Because of the degeneracy of the genetic code, where multiple codons specify the same amino acid, some mutations result in no change in the amino acid sequence and hence are referred to as **samesense** or **silent mutations**. These mutations result in a different base in the third position of a codon, but the new codon specifies the same amino acid (Figure 4.1B). For example, four mRNA codons specify the amino acid valine, GUU, GUC, GUA, and GUG. If the sequence CAG in the DNA, which is transcribed into GUC in the mRNA and translated into valine, is mutated such that the triplet in the DNA becomes CAC, the mRNA codon would be GUG. The codon GUG also specifies valine. Therefore, samesense mutations result in no observable change in the amino acid composition of a protein.

Nonsense mutations are almost inevitably disruptive to protein function because they result in the premature formation of one of the three stop codons, UAG, UGA, or UAA, in the mRNA (Figure 4.1C). Nonsense mutations result in proteins that are shorter than the normal (wild-type) proteins and these "truncated" proteins are generally nonfunctional.

FRAMESHIFT MUTATIONS CHANGE THE IDENTITY OF MULTIPLE CODONS

As mentioned in Chapter 3, the genetic code is translated by reading sequential groups of three bases called codons, beginning from the initiator codon (AUG). The order in which the bases are grouped into codons (which is set by the initiator codon) is called the **reading frame**. If a mutation caused a single base pair to be inserted into a gene sequence, this would have the effect of completely disrupting the reading frame, such that most codons after the point of insertion would be changed (**Figure 4.2**). A similar effect would be achieved by the deletion of a single base pair. Mutations that alter the reading frame in this way are called **frameshift mutations**, and they usually result in nonfunctional proteins that are shorter than the normal protein products. Thus, downstream of the site of insertion or deletion, out-of-frame codons are read until translation is usually terminated prematurely by the substitution of a stop codon for what was originally a codon specifying an amino acid (Figure 4.2).

Because the mRNA sequence is translated by reading consecutive groups of three bases, it follows that an *in-frame* insertion of three bases would not alter the reading frame beyond causing the insertion of an extra codon into the mRNA sequence. Such frameshift mutations can sometimes be tolerated depending on the properties of the added amino acid and where it is inserted in the resulting protein. Similarly, an in-frame deletion of three bases would result in the loss of

-ATG ACC GAC GAG ATG AAT GAA CGT AGC TTT CTG ACG- **DNA**

-AUG ACC GAC GAG AUG AAU GAA CGU AGC UUU CUG ACG- **mRNA**

Met Thr Asp Glu Met Asn Glu Arg Ser Phe Leu Thr- **Amino acid sequence**

↓ **Insertion of one extra Cytosine (C)**

-ATG ACC CGA CGA GA<u>T</u> GAA TGA ACG TAG CTT TCT GAC G- **DNA**

-AUG ACC CGA CGA GA<u>U</u> GAA UGA ACG UAG CUU UCU GAC G- **mRNA**

Met Thr Arg Arg Asp Glu STOP **Amino acid sequence**

↓ **Restoration of reading frame by second-site mutation**

-ATG ACC CGA CGA GAG AAT GAA CGT AGC TTT CTG ACG- **DNA**

-AUG ACC CGA CGA GAG AAU GAA CGU AGC UUU CUG ACG- **mRNA**

Met Thr Arg Arg Glu Asn Glu Arg Ser Phe Leu Thr- **Amino acid sequence**

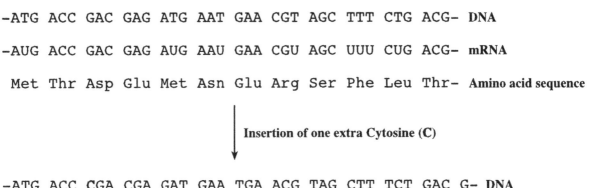

The top three lines indicate wild-type DNA with its mRNA and corresponding amino acid sequences, respectively. If the DNA is subjected to a frameshift mutation such as that caused by the insertion of one extra base, the reading frame would be altered, as shown in the second set of DNA, mRNA, and amino acid sequences. The highlighted cytosine indicates the incorrectly inserted base. The reading frame is shifted such that incorrect amino acids are specified until UGA is reached in the mutant sequence, specifying the termination of protein synthesis. This leads to a prematurely terminated protein. Such a frameshift can be reversed, or reverted, by the deletion of a single base in close proximity to the initial insertion. The underlined thymine (uracil in the mRNA) indicates the deleted base in the third set of sequences. This second-site mutation restores the reading frame downstream of the deletion, leaving only the amino acids between the two mutations affected. If these amino acids are not critical for the function of an active site of an enzyme or for proper protein folding, the protein may regain at least partial function.

Figure 4.2 Frameshift mutations.

a codon from an otherwise unaffected reading frame. The consequences of deleting a codon, and hence an amino acid, will depend on how critical that amino acid is to the function of the protein. For example, one of the most common causes of the lung disease cystic fibrosis, a recessive genetic condition that affects 1 in 2,000 whites, involves the in-frame deletion of the trinucleotide TTT in the gene encoding the cystic fibrosis transmembrane conductance regulator, which is a membrane protein involved in transporting ions across epithelial surfaces, such as the linings of the lungs. Unfortunately, the phenylalanine (Phe) residue that is lost as a result of the in-frame TTT deletion forms part of an ATP-binding site that is critical for the function of this protein. Therefore, despite the fact that only a single amino acid is affected by this in-frame mutation, the consequences for the protein and hence for the affected individual are disastrous.

CHROMOSOMAL MUTATIONS

Chromosomal mutations are grouped into four broad types: **deletions**, **duplications**, **inversions**, and **translocations**. **Deletions** are chromosomal changes in which one or more genes or segments of chromosomal DNA are lost (**Figure 4.3A**). Like frameshift mutations, most chromosomal deletion mutations are lethal unless they affect inessential genes. **Duplications** are chromosomal changes in which one or more copies of a gene are present on the same chromosome. Deletions and duplications can occur simultaneously when two homologous chromosomes break at the same time at two different (nonhomologous) positions and then reconnect to the wrong partner (Figure 4.3B). One of the two chromosomes produced by this mechanism will be missing one or more genes (it would have a deletion) and the other homologous strand would have an extra copy of the gene or genes that were deleted from the first chromosome (it would have a duplication). The breaking and rejoining of a chromosomal molecule can also lead to an **inversion** in which a segment of DNA is released and rotated 180 degrees before being reinserted into the DNA (Figure 4.3C). If the "inverted" DNA segment carries part of a protein-coding sequence, the resulting protein would be drastically altered and most likely nonfunctional.

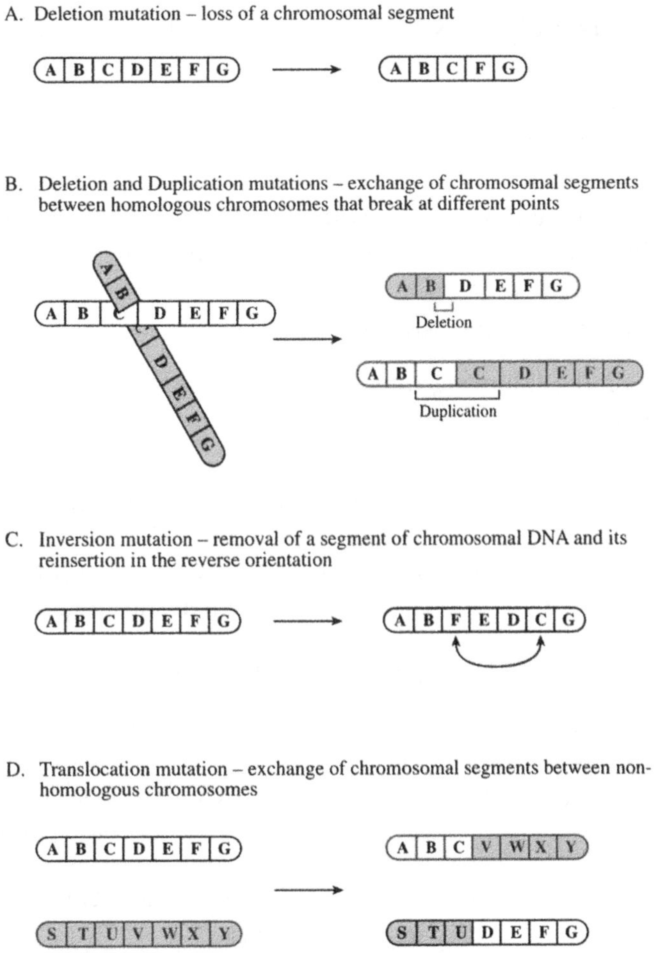

Figure 4.3 Different types of chromosomal mutations. A. Deletion mutation—loss of a chromosomal segment. B. Deletion and duplication mutations—exchange of chromosomal segments between homologous chromosomes that break at different points. C. Inversion mutation—removal of a segment of chromosomal DNA and its reinsertion in the reverse orientation. D. Translocation mutation—exchange of chromosomal segments between nonhomologous chromosomes.

Translocations occur when a segment of DNA moves from one chromosome and is inserted into a different nonhomologous chromosome. Translocations can also be reciprocal, that is, two nonhomologous chromosomes may break and trade pieces of DNA (Figure 4.3D). Translocation mutations frequently cause problems in meiosis and sometimes lead to aneuploidy (the gain or loss of chromosomes).

Topic Test 1: Types of Gene Mutations

True/False

1. One of the most common causes of cystic fibrosis involves a deletion mutation that affects only one codon in the reading frame.
2. Because chromosomal mutations affect large segments of chromosomal DNA, these mutations are always detrimental to the organism.

Multiple Choice

3. An error occurs during DNA replication in a cell, so that where there is supposed to be a C in the middle of one of the genes there is instead an A. Which of the following mutant proteins would be the least likely result of this mutation?
   a. A truncated protein.
   b. An inactive (nonfunctional) protein with one different amino acid.
   c. An active (functional) protein with one different amino acid.
   d. A protein that contains exactly the same amino acids as the protein made from the unmutated (wild-type) gene.
   e. A protein with a single extra amino acid.
4. Which one of the following mutations would be *most likely* to have a harmful effect on the gene product?

Wild type ATG AGG TCT CGC GCA GAG TTG CGA ATT

A ATG AGG TCT CGG GCA GAG TTG CGA ATT
base-pair substitution

B ATG AGG TCT CGC △ GAG TTG CGA ATT
deletion of 3 bases

C ATG AGG TCT CGC GCA GAG GTG CGA ATT
base-pair substitution

D ATG ATG GTC TCG CGC AGA GTT GCG AAT T
insertion of 1 base

5. A mutational event results in the insertion of an extra base pair into the middle of a gene sequence. Which of the following would you expect to be the *least likely* consequence of such a mutation?
   a. A truncated protein product.

b. A protein product in which one amino acid is changed.
c. A protein in which most amino acids after the site of insertion are changed.
d. An inactive (nonfunctional) protein product.
e. All of the above

Short Answer

6. Name, and briefly contrast, the two types of point mutations.
7. A bacterial messenger RNA, AUGCUCUUAGGACUU, codes for fMet-Leu-Leu-Gly-Leu. Use the codon table below to determine what type of mutation would occur if
   a. the second A was replaced by a C
   b. the second G was replaced by a U

Second Position

First Position	U	C	A	G	Third Position
U	Phe	Ser	Tyr	Cys	U
	Phe	Ser	Tyr	Cys	C
	Leu	Ser	Stop	Stop	A
	Leu	Ser	Stop	Trp	G
C	Leu	Pro	His	Arg	U
	Leu	Pro	His	Arg	C
	Leu	Pro	Gln	Arg	A
	Leu	Pro	Gln	Arg	G
A	Ile	Thr	Asn	Ser	U
	Ile	Thr	Asn	Ser	C
	Ile	Thr	Lys	Arg	A
	Met	Thr	Lys	Arg	G
G	Val	Ala	Asp	Gly	U
	Val	Ala	Asp	Gly	C
	Val	Ala	Glu	Gly	A
	Val	Ala	Glu	Gly	G

Topic Test 1: Answers

1. **True.** One of the most common causes of cystic fibrosis involves the in-frame deletion of the trinucleotide TTT. Because a single codon's worth of DNA is removed, the reading frame is maintained, and only one amino acid (a phenylalanine) is lost.
2. **False.** Eukaryotic genomes contain a large amount of noncoding DNA, so if, for example, a chromosomal deletion mutation occurs in a region of DNA that is not transcribed, it may go undetected by the organism. Likewise, a mutational event that occurs in a cell-specific gene will not be detected if it takes place in a cell type in which the gene is not expressed.
3. **e.** If this change from a C to an A changes a codon that specifies an amino acid to one of the three stop codons (e.g., UAC → UAA), the protein would be prematurely terminated

at that point. This means that a truncated protein is possible, and so a is not the answer. If the mutation changes the codon to a different codon that specifies the same amino acid (samesense), the protein sequence would not be changed. So d describes a legitimate possibility and is not the answer. If the mutation changes the amino acid specified (a missense mutation), two possibilities exist. If, for example, the mutation changes one hydrophobic amino acid buried in the interior of a protein to a different hydrophobic amino acid, the mutation is unlikely to affect the function of the protein. This is the situation described in c. On the other hand, if the mutation changes a positively charged amino acid in the active site of an enzyme to a negatively charged amino acid, it is highly likely that the enzyme will become inactive because amino acids in the active site are usually critical for enzyme activity. For this reason, b is not the answer. This leaves option e. It is highly unlikely that the type of mutation described would lead to an increase in the number of amino acids. Therefore, e is the correct answer.

4. **d.** All mutations listed have the potential to be harmful, but as a general rule, the mutation that disrupts the most codons will produce the most severe results. The insertion of one base early in the reading frame, as shown in d, would alter the reading frame from that point to the end of the protein, resulting in multiple amino acid substitutions. Because the base pair substitution in a occurs in the third position of a trinucleotide, it is possible that the resulting codon will specify the *same* amino acid as the original codon and hence that the mutation will be invisible to the organism (a samesense or silent mutation). The base pair substitution in c occurs in the first position of a trinucleotide so it would most likely lead to an amino acid substitution in the resulting protein. However, if the substituted amino acid has similar properties to the original amino acid, it may be tolerated by the protein, which may retain full or partial function. The deletion of three bases in b will remove one amino acid but will not alter the reading frame. If the missing amino acid is not critical for protein function, this mutation may not be that harmful.

5. **b.** An insertion of one base pair in the middle of a gene sequence would cause a frameshift event that would affect every codon (in the resulting mRNA) from the point of insertion to the 3′ end of the gene. So c is not the correct answer, because it describes exactly what *would* happen. Furthermore, because frameshift mutations that change the reading frame usually affect multiple codons, it is highly probable that a codon that formerly specified an amino acid would be changed to one of the stop codons (a **nonsense mutation**), resulting in a truncated protein product. Therefore, a is not the least likely consequence. Likewise, it is probable that even if the protein is not terminated prematurely, it will be composed of a large number of incorrect amino acids when compared with the wild-type protein and will be inactive. So d is also a possible occurrence and therefore not the least likely consequence. This leaves b as the correct answer. Point mutations can lead to changes in single amino acids, but it is highly unlikely that a frameshift mutation will affect only a single amino acid.

6. The two types of point mutations are base substitution mutations and frameshift mutations. Base substitution mutations change a single codon by substituting one base for another—they do not affect the reading frame of the gene, although they can truncate it if a codon specifying an amino acid gets changed to a stop codon. Frameshift mutations involving the insertion or deletion of one or two bases alter the reading frame of the gene from the point of the mutation to the end of the gene. Such frameshift mutations usually cause multiple amino acid substitutions in the protein product. Frameshift mutations

involving the "in-frame" insertion or deletion of three bases would result in a protein product containing one extra or one less amino acid, respectively, but the rest of the reading frame and hence the amino acid sequence would not be affected.

7. **a.** The new mRNA sequence if the second A was replaced by a C would be AUG-CUC-UUC-GGA-CUU. Looking up these codons in the codon table, we see that the new protein sequence would be fMet-Leu-Phe-Gly-Leu. Therefore, the substitution of the second A with a C would cause a **missense mutation**. b. The new mRNA sequence if the second G was replaced by a U would be AUG-CUC-UUA-UGA-CUU. Looking up these codons in the codon table, we see that the new protein sequence would be fMet-Leu-Leu-Stop. Therefore, the substitution of the second G with a U would cause a **nonsense mutation** and would result in a truncated protein product.

TOPIC 2: CAUSES OF MUTATIONS AND DNA REPAIR MECHANISMS

KEY POINTS

✓ *What is the molecular basis of mutations?*

✓ *When do most mutations occur?*

✓ *Why is the number of observed mutations much smaller than the number that actually occur?*

✓ *Why does the repair of DNA damage caused by ultraviolet radiation often cause mutations?*

Mutations occur as a result of **spontaneous** changes in the cell or as a result of exposure to **mutagens**, external agents such as chemicals or physical agents such as ultraviolet (UV) radiation.

INSTABILITY OF THE DNA BASES LEADS TO SPONTANEOUS MUTATIONS

Most spontaneous mutations are the result of the instability of the nucleotide bases in DNA. Thus, all four DNA bases—A, G, C, and T—can exist in an alternative structural form known as a **tautomer**, caused by the spontaneous movement of a hydrogen atom from one position to another within the base. A problem arises because the tautomeric forms of the bases exhibit different base-pairing properties. For example, thymine and guanine are normally in **keto** forms, and in this form, thymine base pairs to adenine and guanine base pairs to cytosine. However, when in their rare **enol** forms, thymine and guanine instead form three hydrogen bonds with the keto forms of guanine and thymine, respectively. Likewise, adenine and cytosine are normally in **amino** forms, but when in their rare **imino** forms, they can form two hydrogen bonds with the amino forms of cytosine and adenine, respectively (**Figure 4.4**). Thus, if a base happens to be in its rare tautomeric form just as it is being replicated by DNA polymerase, an *incorrect* base will be *correctly* (for the tautomeric form of the base) hydrogen bonded to the template strand. If the tautomer in the template strand resumes its normal form after replication, a mismatched base pair will result (Figure 4.4).

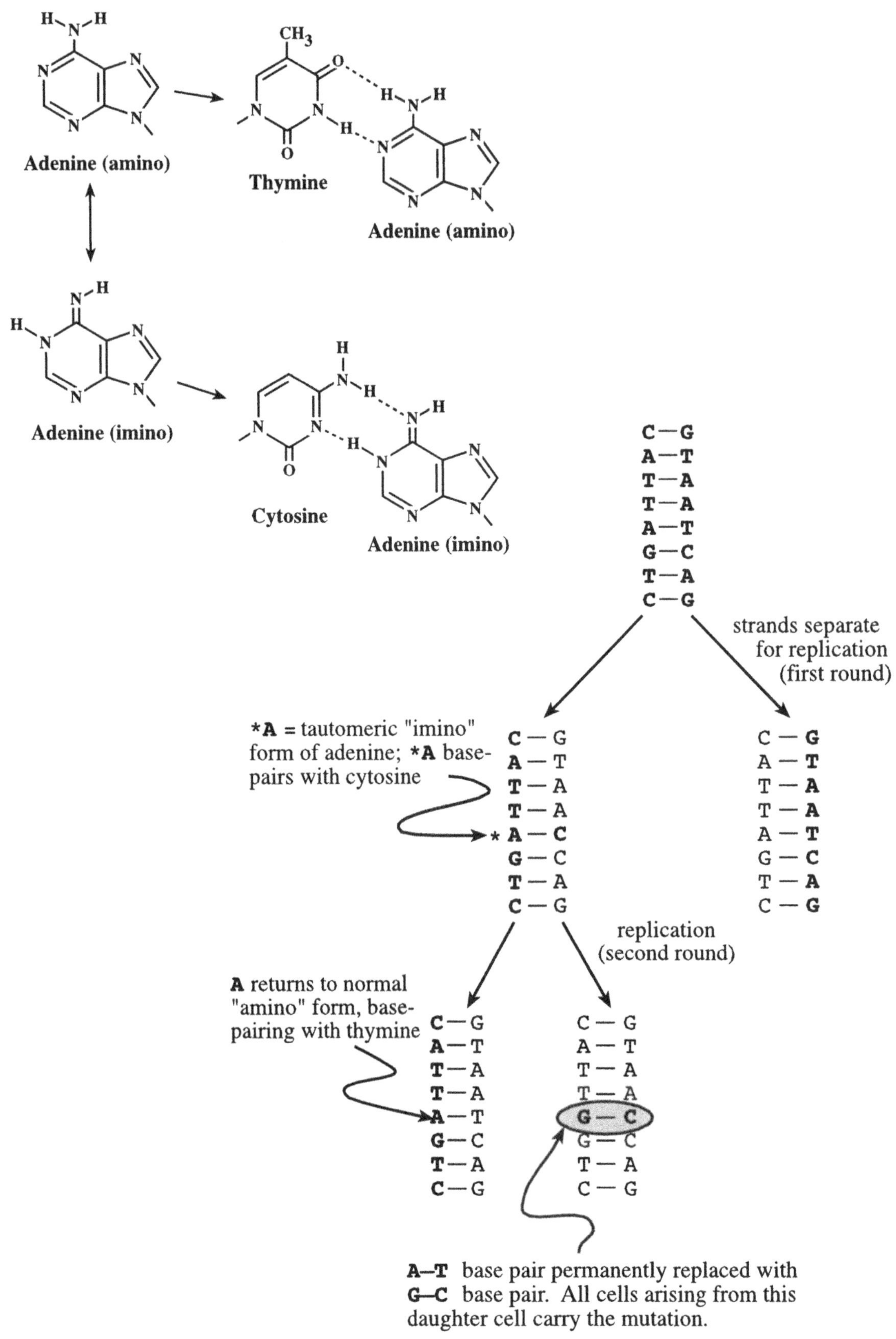

Figure 4.4 Long-term effects of tautomerization.

Another source of mutations is caused by the spontaneous alteration of one of the DNA bases, usually by the loss of a side group. For example, cytosine has a slight tendency to lose its amino group, a process called **deamination**. Because the product of cytosine deamination, uracil, base pairs with adenine instead of guanine, replication of a strand containing a GU base pair ultimately leads to substitution of an AT base pair for the original GC base pair (by the process GC → GU → AU → AT). Adenine can also undergo deamination to form hypoxanthine, which base pairs with cytosine instead of thymine. In this case, the ultimate outcome is the substitution of a GC base pair for the original AT base pair (by the route AT → hypoxanthine C → GC).

Base substitutions resulting from tautomeric shifts or the spontaneous deamination of a cytosine base cause **transitions**. **Transitions** result when a purine is replaced by another purine (A ↔ G) or a pyrimidine is replaced by another pyrimidine (T ↔ C). The second type of base substitution mutations, **tranversions**, occur when the mispairing caused by the formation of an alternative base results in a purine being replaced by a pyrimidine or vice versa (A ↔ C).

MUTATIONS CAN BE INDUCED BY CHEMICAL OR PHYSICAL AGENTS

Although mutations occur spontaneously, they occur much more frequently in response to environmental agents. Chemical and physical agents that cause mutations are called **mutagens**. Mutagens can cause mutations by

1. replacing one base with a so-called **base analogue** with different base-pairing properties;
2. modifying a previously inserted base to a base with different base-pairing properties (e.g., **alkylating agents**);
3. causing the insertion or deletion of one or a few bases (e.g., **intercalating agents**);
4. damaging a base so that it can no longer form hydrogen bonds with the complementary strand (e.g., **radiation**).

HIGH RATE OF TAUTOMERIZATION OF BASE ANALOGUES LEADS TO BASE PAIR SUBSTITUTIONS

A **base analogue** is sufficiently similar to one of the four DNA bases that it can be incorporated into a replicating DNA molecule by DNA polymerase. For example, 5-bromouracil (BU) is identical to thymine except for the substitution of a bromine atom for a methyl group, and like thymine, in its keto form BU can form two hydrogen bonds with adenine. As discussed previously, thymine occasionally exists in a rare enol form that base pairs with guanine instead of adenine. The mutagenic activity of BU stems in part from a shift in the keto-enol equilibrium caused by the bromine atom, such that conversion to the enol form of BU occurs at a much higher rate than for thymine. Thus, if BU replaces a thymine in a replicating DNA molecule, it is highly likely that in subsequent rounds of replication, the BU will shift into its enol form and will be base paired with guanine instead of adenine (**Figure 4.5**). The misincorporated guanine will ultimately result in the substitution of a GC base pair for the original AT base pair.

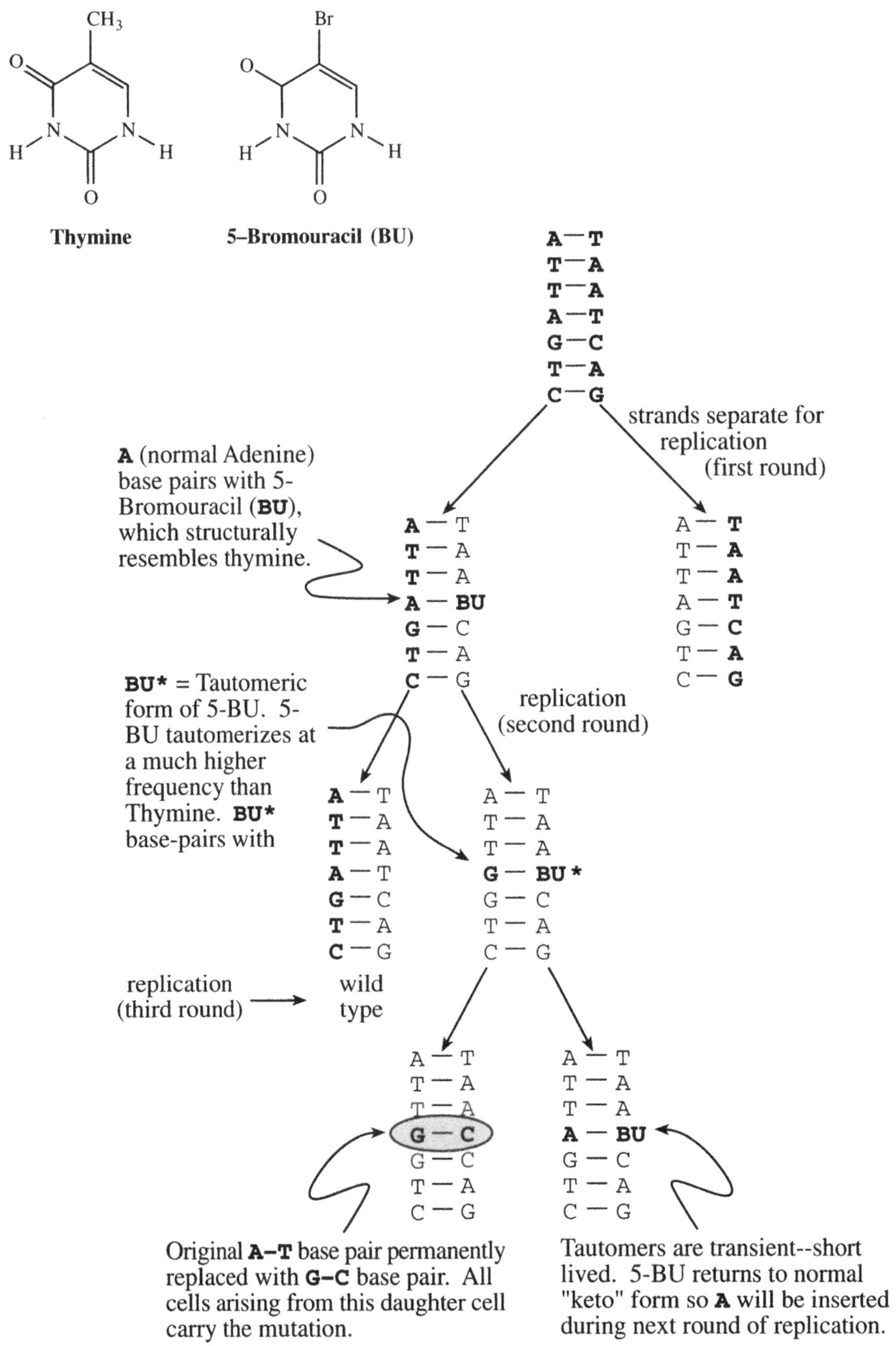

Figure 4.5 Long-term effects of 5-bromouracil incorporation.

MODIFICATION OF DNA BASES ALTERS THEIR BASE-PAIRING PROPERTIES

Alkylating agents cause mutations by chemically altering one of the DNA bases in such a way that it can no longer form hydrogen bonds with its normal complementary base. For example, the alkylating agent ethylmethane sulfonate (EMS) adds an alkyl group to the hydrogen-bonding oxygen of guanine. The resulting alkylated guanine can no longer base pair to cytosine but instead forms hydrogen bonds with thymine. This ultimately results in the substitution of an AT base pair for the original GC base pair.

INTERCALATION OF PURINE-PYRIMIDINE BASE-PAIR ANALOGUES CAUSES FRAMESHIFT MUTATIONS

Intercalating agents such as ethidium bromide and acridine orange are planar heterocyclic molecules with dimensions approximately the same as those of a purine-pyrimidine base pair. Because of their similarity to a normal Watson-Crick base pair, intercalating molecules can insert themselves between the stacked nucleotide bases within the DNA double helix, a process called **intercalation**. When DNA containing an intercalating agent is replicated, one or more additional bases often appear in the sequence of the new DNA strand or, occasionally, a single base may be deleted. Because the base sequence is read in groups of three bases during translation into an amino acid sequence, the addition or deletion of one or two bases changes the reading frame. Thus, intercalating agents cause frameshift mutations; downstream of the site of intercalation, out-of-frame codons will be read until translation is usually terminated prematurely by the conversion of an amino acid-specifying codon to a stop codon.

RADIATION DAMAGES NUCLEOTIDES

When **UV light** is absorbed by adjacent thymines in one strand of a double-stranded DNA molecule, the thymines fuse to form a single covalently linked thymine dimer. At the site of a thymine dimer, the DNA double helix is distorted, and in *E. coli* this distorted helical structure prevents DNA polymerase III from replicating the DNA opposite the dimer. The impact on DNA replication is so extensive that cell division is stopped until the dimers are repaired or bypassed. One of the mechanisms that repairs UV damage in *E. coli* cells, called the **SOS-repair mechanism**, induces **error-prone** replication. The SOS-repair mechanism is an emergency repair process aimed at bypassing "replication blocks" such as those caused by the formation of thymine dimers. The mechanism by which the SOS proteins repair the DNA is not fully understood, but it is thought that the SOS proteins interact with DNA polymerase III and somehow relax the requirement for Watson-Crick base pairing. Thus, when the SOS proteins are induced, stalled DNA polymerases bypass the thymine dimer by incorporating one or more bases at random until the polymerase passes through the damaged site. As a result of the relaxed base pairing, a base other than the normal complementary base often gets added, resulting in a viable cell carrying one or more mutations.

Gamma and x-rays, which are more energetic than UV rays, cause most of their DNA damage indirectly by ionizing the molecules surrounding the DNA, especially water. This ionization leads to the formation of **free radicals**, chemical substances with an unpaired electron, which are predominantly highly reactive oxygen species. When a free radical attacks a DNA molecule, it can cause a single- or double-stranded break. Whereas single-stranded breaks are easily repaired, the repair of double-stranded breaks frequently occurs improperly resulting in a mutation.

REVERSION

Most point mutations are reversible. **Reversion** to the original phenotype can occur in two ways. First, true reversion occurs when the mutated base or base sequence is restored to its original identity. Second, the original phenotype can also be restored by a second mutation occurring elsewhere in the same gene that compensates for the first mutation. For example, a forward shift in the reading frame caused by the insertion of a single base can be corrected by a second frameshift mutation occurring nearby that deletes a base to compensate for the inserted base (Figure 4.2). Alternatively, consider a positively charged arginine residue that is critical for the proper folding of a protein due to its interactions with a negatively charged glutamic acid residue elsewhere in the protein. A mutation that substituted a negatively charged amino acid for the positively charged arginine would inevitably prevent proper folding and lead to loss of activity. However, if a second mutation substituted the negatively charged glutamic acid that the original arginine interacted with, for a positively charged amino acid, proper folding and hence activity might be restored.

MUTAGENS ACCELERATE THE MUTATION RATE, INCLUDING THE RATE OF REVERSIONS

Because environmental mutagens are thought to be responsible for most human cancers, it is important to identify mutagenic chemicals so that exposure to them can be limited. One of the best systems for detecting mutagens was developed by Bruce Ames and is known as the **Ames test**. This test uses auxotrophic histidine mutants of *Salmonella typhimurium* that are unable to grow in the absence of the amino acid histidine. These mutants are also deficient in some of the DNA repair mechanisms. The basic idea of the assay is to use these mutants to observe the frequency of reversion to a wild-type (*his*$^+$) phenotype in the absence and presence of a potential mutagen. This is done very simply by plating the control and treated cells onto medium lacking histidine—only *his*$^+$ revertants will be able to grow. Mutagens greatly accelerate the mutation rate and hence the frequency of reversions so, in the presence of a mutagen, many more *his*$^+$ colonies will be observed. Following the observation that many chemicals become mutagenic only after they have been acted upon by liver enzymes, Ames modified his test by growing the cells in the presence of rat liver extract to mimic the metabolic changes a potential mutagen would experience in the human body. Using this test, many potentially mutagenic chemicals have been identified.

ENZYMES CORRECT MANY OF THE MISTAKES MADE BY DNA POLYMERASES

Most mutations occur during DNA replication when huge amounts of new DNA are being synthesized. Because errors in the genetic material can have fatal consequences, many DNA polymerases (both DNA polymerases I and III in *E. coli* and DNA polymerase δ in eukaryotes) possess a **3′ to 5′ exonuclease** or **proofreading** activity. This activity enables DNA polymerases to remove misincorporated nucleotides. Thus, when *E. coli* DNA polymerase III adds the wrong nucleotide to a growing chain, abnormal structural changes occur in the DNA molecule that can be recognized by one of the subunits of the DNA polymerase III enzyme. The DNA polymerase can then remove the incorrect nucleotide, leaving a free 3′-OH group for the enzyme to try again. This proofreading activity greatly reduces the final error rate. Thus, in *E. coli*, it is estimated that during DNA replication, DNA polymerase III makes one mistake for every 10^4 base pairs (this translates to one mistake in approximately 1 of every 10 genes each time the cell divides). The proofreading activity of the *E. coli* DNA polymerases reduces this error rate by approximately 10^2 but this still leaves an error rate of about one for every 10^6 base pairs.

The error rate is reduced even further by a second repair mechanism. Thus, immediately after a piece of DNA has been replicated, **mismatch repair enzymes** survey the new double-stranded DNA molecule looking for mismatched bases. Once located, the mismatch repair enzymes remove the mismatched base and a number of adjacent bases from the newly synthesized DNA strand, leaving a stretch of single-stranded DNA that DNA polymerase can reuse as a template for DNA replication. The mismatch repair mechanism reduces the error frequency by another 10^2, accounting for the observed error frequency in *E. coli* of one mistake for every 10^8 base pairs ($10^4 \times 10^{-2} \times 10^{-2}$). How do the mismatch repair enzymes know which strand to repair? For example, suppose the mismatch repair system detects a mismatched GT base pair: How does the repair mechanism know which base to remove (i.e., how do the repair enzymes know whether the G or the T is the new, and hence presumably the incorrect, base)? In *E. coli*, the DNA is heavily methylated. However, after DNA replication there is a lag before the newly synthesized DNA is completely methylated. Thus, the mismatch repair enzymes can distinguish new from "old" DNA because newly synthesized DNA is undermethylated for a short period of time compared with the parental DNA. The unmethylated DNA strand that is present in a double-stranded DNA molecule immediately after replication is the newly synthesized DNA and hence is recognized by the repair enzymes as the strand with the errors.

OTHER MECHANISMS REPAIR ERRORS THAT OCCUR DURING THE LIFE OF A CELL

In **excision repair**, enzymes recognize mismatched base pairs, chemically modified bases, or base additions or deletions in a cell's DNA. These enzymes can remove damaged or altered bases in the DNA, leaving single-stranded template for DNA polymerase to repair.

In so-called **recombination repair**, the DNA damage is not actually repaired but instead is bypassed until it can be repaired at a later time. In this mechanism, replication occurs on either side of the damaged DNA, and the gap that remains in the newly synthesized DNA strand is filled in by recombination with its homologous strand in the other daughter DNA molecule. This recombination event fills in the gap opposite the damaged DNA but creates a new gap in the other daughter DNA molecule. However, because this daughter DNA molecule has no DNA damage, the new gap created by the recombination event can easily be filled in by DNA polymerase and DNA ligase.

Photoreactivation or **light repair** is catalyzed by the enzyme **DNA photolyase** and is the most important mechanism for repairing the damage caused by UV light in bacteria and in many lower eukaryotes (DNA photolyase is not thought to be present in most higher eukaryotes). DNA photolyase can detect pyrimidine dimers caused by UV damage, and in the presence of visible light that activates it, DNA photolyase can break the covalent bond holding the pyrimidine dimer together.

Like recombination repair, the **error-prone** or **SOS-repair mechanism** is another mechanism that allows the cell to bypass DNA damage without actually repairing it. As discussed above, the error-prone repair enzymes appear to interact with *E. coli* DNA polymerase III and allow it to form non–Watson-Crick base pairs. As a result, as the DNA polymerase III replicates across the pyrimidine dimer, a base other than the normal complementary base often gets added, resulting in a viable cell carrying one or more mutations.

Topic Test 2: Causes of Mutations and DNA Repair Mechanisms

True/False

1. If a large number of base pair substitution mutations occurred that resulted in GC base pairs being replaced by AT base pairs, you could conclude that the DNA had been exposed to BU.
2. The Ames test identifies potential mutagens based on their ability to knock out histidine biosynthesis in *S. typhimurium* mutants.
3. Certain DNA repair mechanisms appear to be based on the principle that a few mutations are better than no DNA replication at all.
4. All four DNA bases can exist in alternative tautomeric forms: Thymine and guanine can alternate between keto and enol forms, adenine and cytosine between amino and imino forms.

Multiple Choice

5. Which one of the following mutagen types would you expect to cause the *greatest* disruption to a DNA sequence?
   a. A base analogue
   b. An alkylating agent
   c. An intercalating agent
   d. UV radiation
   e. a and b
6. Which one of the following types of mutagen is the *most likely* to cause a base substitution mutation by incorporating into a replicating DNA molecule?
   a. A base analogue
   b. An alkylating agent
   c. An intercalating agent
   d. UV radiation
   e. All of the above

Short Answer

7. A mutant strain was treated with the following mutagens to see if revertants could be produced: BU (a base analogue), EMS (an alkylating agent), and ICR-191 (an intercalating agent). In the following table, + = revertants and – = no revertants. Determine the probable type of mutation that occurred to produce this mutant strain.

Chemical		
5-bromouracil	EMS	ICR-191
-	-	+

8. Describe how UV light causes damage to DNA and how some of the enzymes used to repair this damage can actually cause mutations.

9. In the Ames test, why are cells grown in the presence of rat liver extract?

Topic Test 2: Answers

1. **False.** BU can substitute for thymine. When BU shifts into its enol form (which it does at a much higher frequency than thymine), it base pairs with guanine instead of adenine. The misincorporated guanine would ultimately result in the substitution of a GC base pair for the original AT base pair, not the other way around.

2. **False.** The Ames test identifies potential mutagens on the basis of their ability to revert *his*$^-$ *S. typhiumurium* mutant strains to a *his*$^+$ phenotype.

3. **True.** An example of this is the "error-prone" SOS-repair system. If a cell is unable to replicate its entire genome, the cell will die. On the other hand, a small number of mutations, such as those introduced by SOS-repair across a replication block, may not be lethal to the cell.

4. **True.** These temporary transitions to rare tautomeric forms lead to alternative base pairing and are a common cause of base substitution mutations.

5. **c.** The greatest disruption to a DNA sequence is caused by frameshift mutations because they alter most of the reading frame (and hence most of the codons) from the point of the mutation to the end of the gene. Base analogues, alkylating agents, and UV radiation all typically give rise to point mutations, where existing base pairs are altered, but new base pairs are not added nor are existing base pairs removed. Intercalating agents, on the other hand, because of their structural similarity to a normal Watson-Crick base pair, can insert themselves between the stacked nucleotide bases in a DNA double helix. To compensate for the extra space occupied by the intercalating agent, when DNA containing an intercalating agent is replicated, one or more additional bases often appear in the sequence of the new DNA strand or, occasionally, a single base may be deleted, resulting in a frameshift mutation.

6. **a.** On the surface, it may seem like there are multiple correct answers to this question. c is not correct, however, because intercalating agents typically give rise to frameshift mutations rather than base substitution mutations. This also rules out e as a possibility. However, of the remaining options, only a is the correct answer for the *question asked.* The

question asks which of the mutagens can cause base substitution mutations by *incorporating* into the replicating DNA molecule, not which of the mutagens simply leads to substitution mutations. UV radiation causes the formation of thymine dimers, but the UV radiation itself is not an entity that can incorporate into the DNA. Likewise, EMS changes a guanine so that it can no longer base pair with cytosine, but the EMS molecule itself does not become part of the DNA (it alters existing DNA bases and, in so doing, changes their base-pairing properties). Only a base analogue such as BU can actually integrate into the replicating DNA molecule where it substitutes for thymine.

7. Both BU and EMS cause mutations that result in base substitutions. Because reversions are simply second mutations that reverse or compensate for the effects of the original mutations, it follows that BU and EMS can reverse base substitution mutations. However, neither of them reverts the mutant strain, suggesting that the mutant strain did not arise from a base substitution mutation. ICR-191 is an intercalating agent and would be expected to revert a frameshift mutation. For example, if the original mutation resulted in the deletion of a single base, ICR-191 could cause a singe base to be inserted near the original mutation, thereby restoring the original reading frame. Because treatment with ICR-191 did result in an increase in the number of revertants, the mutant strain most likely arose from a frameshift mutation.
8. UV radiation causes adjacent thymines to form a covalent dimer, distorting the double helix and preventing DNA polymerase from replicating the DNA in that region. In *E. coli*, if the photoreactivation, excision repair, or recombination repair mechanisms fail to repair UV damage, the SOS-repair mechanism is called into action. The SOS-repair mechanism attempts to save the cell by allowing "error-prone" replication across from the thymine dimers. In *E. coli*, it is thought that the SOS proteins interact with DNA polymerase III and somehow relax the requirement for Watson-Crick base pairing. Thus, when the SOS proteins are induced, stalled DNA polymerases bypass the thymine dimer by incorporating one or more bases at random until the polymerase passes through the damaged site. As a result of the relaxed base pairing, a base other than the normal complementary base often gets added, resulting in a viable cell carrying one or more mutations.
9. It was observed that many potential mutagens only become mutagenic after they have been acted upon by liver enzymes. Therefore, if the Ames test is performed in the absence of liver enzymes, many chemicals will appear to be nonmutagenic when in fact they would be converted into mutagens in the body.

IN THE CLINIC

The analysis of mutant phenotypes has provided an enormous amount of insight into virtually all aspects of molecular biology. For example, it is largely through the analysis of mutants of both bacteria and phages that many fundamental molecular processes in the living cell were characterized. Similar approaches have subsequently been used to identify two types of genes associated with various cancers, **oncogenes** and **tumor suppressor genes**.

Proto-oncogenes such as the ***ras* gene**, which is mutated in about 30% of human cancers, code for proteins that stimulate normal cell growth and cell division. Oncogenes arise from mutations in proto-oncogenes that lead to an increase in either the amount or the intrinsic activity of the proto-oncogene protein product. For example, many of the *ras* oncogenes have point mutations that lead to hyperactive versions of the **Ras protein**. Because the role of the Ras protein is to induce the synthesis of a protein that stimulates the cell cycle, a hyperactive version of the Ras protein leads to excessive cell division, the hallmark of cancer.

In contrast to proto-oncogenes, tumor suppressor genes such as the ***p53* gene**, which is mutated in about 50% of human cancers, give rise to proteins that act to suppress the cell cycle. For example, damage to a cell's DNA leads to the production of the **p53 protein**, which itself acts as a transcription factor (see Chapters 2 and 6) for several genes. p53 activates genes directly involved in DNA repair and a gene whose product halts the cell cycle, allowing time for the cell to repair the DNA. Finally, if the DNA damage is irreparable, p53 activates genes whose products cause cell death. Using these three strategies, p53 prevents a cell from passing on mutations, and mutations that knock out the *p53* gene usually lead to excessive cell growth and cancer. Interestingly, the recently discovered breast cancer genes, *BRCA1* and *BRCA2*, are also considered to be tumor suppressor genes because their wild-type alleles protect against breast cancer.

Understanding the genetic basis of these and other genes associated with inherited diseases may lead to the development of new methods for the early diagnosis and treatment of these diseases.

DEMONSTRATION PROBLEM

A wild-type gene has the following amino acid sequence:

Met—Val—Ile—Ala—Pro—Trp—Ser—Glu—Lys—Cys—His—

You recover several mutant versions of this gene with the following sequences:

Mutant 1: Met—Val—Ile—Ala—Pro—Trp—Arg—Glu—Lys—Cys—His—
Mutant 2: Met—Val—Ile—Ala—Pro
Mutant 3: Met—Val—Ile—Ala—Pro—Gly—Val—Lys—Asn—Val—Ile—
Mutant 4: Met—Val—Lys—Cys—His—

Which type of mutation is likely to have produced each mutant version?

Mutant 1, compared with the wild-type sequence, differs in one amino acid near the middle of the gene (conversion of a serine to an arginine). This suggests a **missense mutation** because only one amino acid is changed—all amino acids that follow the substituted amino acid (Glu—Lys—Cys—His) are wild type. This eliminates the possibility of a frameshift mutation, such as a single base insertion or deletion, because such a mutation would alter the reading frame and affect multiple amino acids. Samesense mutations can also be disregarded, because a samesense mutation would be undetectable in a protein sequence (the codon is different, but the amino acid inserted is the same). For example, suppose the codon for serine in the wild-type gene is UCA. If this codon was changed to UCG by a mutation in the DNA sequence, the specified amino acid would still be serine (a samesense mutation).

Mutant 2 contains wild-type amino acids through proline, then stops. This strongly suggests a **nonsense mutation**. Whenever a truncated protein with otherwise wild-type amino acids is detected, it is most likely from a single base mutation that creates an in-frame stop codon. In this example, the amino acid that follows the proline in the wild-type protein is tryptophan. If the codon for tryptophan in this gene is UGG, it could be easily mutated to UGA, a stop codon, thereby terminating protein synthesis immediately after the proline.

In contrast to mutant 2, which stops after proline, mutant 3 continues past proline but does so with amino acids that are different from those specified in the wild-type protein. Because the mutation affects *all* the amino acids downstream of the mutation, it is not likely to be a missense mutation as with mutant 1. Nor is it a samesense mutation, because the mutation is detectable and multiple amino acids are affected. Instead, mutant 3 was most probably caused by a **frameshift mutation** occurring in the middle of the protein. This mutation altered the reading frame from the point of the mutation downstream. If we examine the changes that occurred in the amino acid sequence, we can see how this could occur. For example, let us speculate that the wild-type sequence of the gene, from the tryptophan codon downstream, is as follows:

UGG	AGU	GAA	AAA	UGU	CAU
Trp	Ser	Glu	Lys	Cys	His

Now, let us assume that the mutation that occurs in the gene is the deletion of the uracil in the tryptophan (Trp) codon and that the base that follows the histidine (His) codon is a cytosine (C). The new sequence would be

GGA GUG AAA AAU GUC AUC

This single base deletion shifts the reading frame such that the resulting protein sequence is as follows:

GGA	GUG	AAA	AAU	GUC	AUC
Gly	Val	Lys	Asn	Val	Ile

This corresponds with the protein sequence following the proline in mutant 3 and strongly suggests that the mutation was caused by the deletion of a single base.

Finally, mutant 4 appears to be missing the amino acids Ile—Ala—Pro—Trp—Ser—Glu, which appear in the wild-type protein between the valine near the beginning and the lysine near the end. All the mutation types discussed so far are point mutations. However, point mutations, whether frameshift mutations or base substitution mutations, cannot account for the loss of multiple amino acids out of the middle of a protein sequence. Only chromosomal mutations can give rise to such a phenotype. Therefore, mutant 4 was most likely caused by a **chromosomal deletion**—maybe arising from a recombination event involving homologous chromosomes joining at nonhomologous positions.

Now let's move on to a different type of problem. Imagine that a protein produced by a wild-type strain of *E. coli* has a valine (Val) residue at position 90. Two single base pair substitution mutants (X and Y) are isolated and found to have glycine (Gly) and glutamic acid (Glu) substitutions at this position. **Functional** revertants of mutants X and Y had, respectively, tryptophan (Trp) and lysine (Lys) residues at position 90. Deduce the three-letter codon present at position 90 in the original wild-type strain.

Functional revertants indicate that regardless of the nucleic acid or amino acid sequence, the mutant protein in the revertant regains the wild-type protein's activity. The possible codons for Val are GUU, GUC, GUA, and GUG. Any of these can give rise to Gly from a single base change, but it is not possible to specify Glu from single base pair changes in either GUU or GUC, which eliminates these as possibilities. If the initial codon was GUA, the single base change that would give rise to a Gly codon would be GUA → GGA. We now need to change GGA to a Trp codon with only a single base. However, this cannot be done, because the only codon for Trp is UGG, which would require two bases to be changed. This leaves GUG as the answer, which we can verify. Starting with GUG as the codon for Val, we can form Gly by the route GUG → GGG. The reversion event that follows is GGG → UGG, yielding a Trp codon. Likewise, mutant Y was produced from the single base change GUG → GAG (Glu), and the subsequent reversion event is GAG → AAG (Lys).

Chapter Test

True/False

1. Proofreading activity is solely responsible for reducing the number of errors made by DNA polymerases.
2. The spontaneous tautomerization of nucleotides during DNA replication can lead to base pair substitution mutations.

Multiple Choice

3. Using the codon table provided previously, determine which of the following amino acid substitutions could occur as a result of a single base change in the methionine (Met) codon.
   a. Asp
   b. Arg
   c. Ala
   d. Ser
   e. Stop
4. Which one of the following statements about mutations is correct?
   a. A base pair substitution mutation can cause protein synthesis to be terminated prematurely.
   b. A single base pair substitution mutation can lead to the substitution of multiple amino acids in a protein product.
   c. Intercalating agents can revert base pair substitution mutations.
   d. The number of observed mutations is normally much higher than the number of mutations that actually occur.
   e. The proofreading properties of DNA polymerases rely on their 5′ to 3′ exonuclease activities.
5. Using the codon table provided previously, determine which one of the following base substitutions would result in a samesense mutation.
   a. UUU → UUG
   b. ACA → AAA
   c. AAU → AGU
   d. CGA → AGA
   e. GUG → GAG

6. Approximately one third of the base substitutions that occur in coding regions of DNA are harmless. Why?
   a. Because the genetic code is degenerate.
   b. Because most codons specify more than one amino acid.
   c. Because all proteins remain fully active with only two thirds of their correct complement of amino acids.
   d. Because most DNA in a gene is intron DNA, which is removed anyway.
   e. Because many base substitutions are corrected by RNA polymerase before the resulting mRNA is translated into a protein.
7. Which one of the following phenotypes is the most likely consequence of a nonsense mutation?
   a. The mutant protein has normal (wild-type) activity.
   b. The mutant protein has a substituted amino acid at the active site.
   c. The mutant protein is much shorter than the normal (wild-type) protein.
   d. The mutant protein is temperature sensitive.
   e. All of the above

Short Answer

8. A wild-type protein has a serine (Ser) residue at a particular position. List the possible amino acid substitutions that could occur as a result of a single base pair substitution in the corresponding DNA.
9. When a mutation leads to the replacement of one codon for another that specifies the same amino acid, the resulting mutation is known as a ________.
10. Match each mutagen (1 through 4) with one of the descriptions given below (a through d).
    1 = BU
    2 = EMS
    3 = ICR-191 (intercalating agent)
    4 = Ultraviolet radiation

    a = formation of thymine dimers in DNA
    b = substitution of thymine during DNA replication
    c = chemical modification of normal DNA bases
    d = normal Watson-Crick base pair analogue

Essay

11. A DNA sequence that codes for a protein is mutated, resulting in the base substitution mutations listed below. Use the codon table provided to figure out what type of mutation each base substitution will cause in the resulting protein. (You should assume that the trinucleotide sequences shown are in the template DNA strand that is transcribed by RNA polymerase into mRNA.)
    a. AAA → AAG
    b. TAA → TAC
    c. GTT → ATT

Second Position

First Position	U	C	A	G	Third Position
U	Phe Phe Leu Leu	Ser Ser Ser Ser	Tyr Tyr Stop Stop	Cys Cys Stop Trp	**U** **C** **A** **G**
C	Leu Leu Leu Leu	Pro Pro Pro Pro	His His Gln Gln	Arg Arg Arg Arg	**U** **C** **A** **G**
A	Ile Ile Ile Met	Thr Thr Thr Thr	Asn Asn Lys Lys	Ser Ser Arg Arg	**U** **C** **A** **G**
G	Val Val Val Val	Ala Ala Ala Ala	Asp Asp Glu Glu	Gly Gly Gly Gly	**U** **C** **A** **G**

Chapter Test Answers

1. **False**
2. **True**
3. **b** 4. **a** 5. **d** 6. **a** 7. **c**
8. Phe, Tyr, Cys, Pro, Thr, Ala, Leu, Trp, Ile, Asn, Arg, Gly
9. Samesense mutation
10. 1 = b, 2 = c, 3 = d, 4 = a
11. **a.** If AAA is the template DNA, UUU is the sequence in the mRNA. UUU codes for phenylalanine (Phe). A mutation that changes AAA to AAG consequently changes UUU to UUC but still gives rise to Phe, which makes mutation a a samesense mutation. **b.** TAA in the template gives rise to AUU in the mRNA. AUU specifies Ile. If TAA is changed to TAC, the mRNA sequence changes from AUU to AUG, the codon for methionine (Met). A change from one amino acid to a different amino acid is a missense mutation. **c.** GTT in the DNA leads to CAA in the mRNA, which specifies Gln. Changing GTT to ATT leads to UAA in the mRNA, which is one of the stop codons. This makes mutation c a nonsense mutation.

Check Your Performance

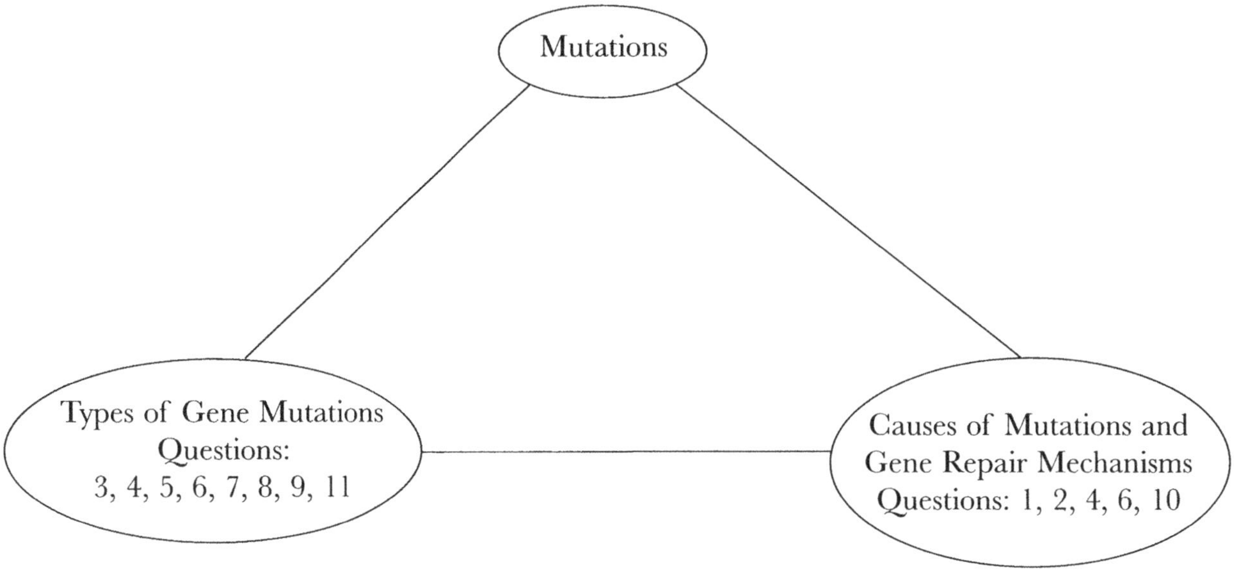

Note the number of questions in each grouping that you got wrong on the chapter test. Identify where you need further review and go back to relevant parts of this chapter.

Midterm Exam

True/False

1. During DNA replication, two new double helices are formed from each parental DNA molecule. These new double helices are hybrid molecules, each individual DNA strand consisting of a mixture of newly synthesized and original parental DNA.
2. The binding of bacterial RNA polymerase requires a subunit of the RNA polymerase holoenzyme that dissociates from the enzyme shortly after initiation of RNA synthesis.
3. In bacteria ribosomes can attach to an mRNA before synthesis of that mRNA is complete.
4. The bonds that form between the anticodon of a tRNA molecule and the codon of an mRNA molecule always involve (form with) the same three bases in the mRNA for any one particular tRNA molecule.
5. One of the frames in an animation of protein synthesis shows a ribosome holding <u>two</u> tRNA molecules. One tRNA has a tetrapeptide (a protein chain four amino acids long) attached to it and it is located in the P site of the ribosome. The other tRNA is attached to a single amino acid and it is located in the A site of the ribosome. The next frame in the animation would show the tetrapeptide leaving its tRNA in the P site and joining to the amino acid attached to the tRNA in the A site.

Multiple Choice

6. Which one of the following statements about the DNA double helix is incorrect?
   a. In a DNA double helix, the number of guanines divided by the number of cytosines = 1.
   b. If a DNA double helix is 20% cytosine, it is 30% adenine.
   c. The DNA double helix consists of two DNA chains that wind around each other in such a way that a purine in one of the chains is always opposite a pyrimidine in the other chain and vice versa.
   d. The three-dimensional conformation of a DNA double helix is strongly influenced by its hydrophobic components.
   e. None of the above—all of the above statements about the DNA double helix are correct.
7. If double-stranded DNA molecules with the following 5′ to 3′ nucleotide sequences were heated, which DNA molecule would you expect to denature at the lowest temperature?
   a. 5′-ATGCTATAGACTACTA-3′ No salt
   b. 5′-GAATATCTATTAGACA-3′ No salt
   c. 5′-GGCCAGCGCCGCG-3′ + salt
   d. 5′-ATGCTATAGACTACTA-3′ + salt
   e. 5′-GAATATCTATTAGACA-3′ + salt
8. Which of the following is/are not normally affected by the treatments used to denature double-stranded DNA molecules?
   a. Base stacking interactions
   b. The hydrogen bonds between the complementary base pairs

c. The phosphodiester linkages in the sugar-phosphate backbone
d. Absorption of UV light
e. All of the above are affected by DNA denaturation.

9. A person's DNA contains information that (directly or indirectly) instructs the cell how to build
a. functional RNAs.
b. DNA.
c. proteins.
d. enzymes.
e. All of the above

10. Which one of the following statements about the differences between DNA and RNA is incorrect?
a. DNA and RNA contain different sugars.
b. DNA and RNA contain different purine bases.
c. DNA usually exists in a double-stranded form; RNA usually exists in a single-stranded form.
d. DNA is usually much larger than RNA.
e. In eukaryotes, DNA is confined to the nucleus or organelles whereas most RNA is not confined to a single cellular compartment.

11. A short piece of DNA has the following nucleotide sequence:

$$5'\text{-}A_1A_2G_3C_4C_5C_6T_7A_8C_9A_{10}A_{11}C_{12}\text{-}3'$$

I = This nucleotide has a triphosphate group
II = This nucleotide has a free 3′ hydroxyl (3′-OH) group
III = The triphosphate group of this nucleotide formed a phosphodiester bond with the 3′-OH group of G_3
IV = The 3′-OH group of this nucleotide formed a phosphodiester bond with the triphosphate group of T_7

Which of the following correctly pairs each nucleotide with its description:
a. I = A_1 II = C_{12} III = C_4 IV = A_8
b. I = C_{12} II = A_1 III = A_2 IV = A_8
c. I = C_{12} II = A_1 III = C_4 IV = C_6
d. I = A_1 II = C_{12} III = A_2 IV = C_6
e. I = A_1 II = C_{12} III = C_4 IV = C_6

12. In bacterial DNA replication, the ________ enzyme contains a 5′ to 3′ polymerase activity, a 5′ to 3′ exonuclease activity, and a 3′ to 5′ exonuclease activity.
a. DNA polymerase I
b. DNA polymerase III
c. DNA polymerase I and DNA polymerase III
d. DNA ligase
e. All of the above

13. What is the sequence of events during DNA replication?
I. Sealing gaps
II. DNA synthesis

III. Unwinding
IV. RNA synthesis
V. RNA replacement

a. III, IV, II, V, I
b. IV, III, II, V, I
c. III, IV, II, I, V
d. III, II, IV, V, I
e. IV, III, II, I, V

14. Which of the following is not involved in the process of DNA replication?
    a. Ribonucleoside triphosphates (ATP, GTP, UTP, CTP)
    b. Enzymes capable of synthesizing RNA
    c. Enzymes capable of removing DNA nucleotides
    d. An enzyme capable of removing RNA nucleotides
    e. All of the above are involved in DNA replication.

15. Choose the option below that provides the correct answers to the following questions on leading strand synthesis during DNA replication in bacteria.
    i) From what substrates is the leading strand of DNA made?
    ii) On what template?
    iii) With what enzyme?
    iv) Is a primer required?

    a. i) = dATP, dGTP, dCTP, dTTP; ii) = single-stranded DNA; iii) = DNA polymerase III; iv) = No
    b. i) = dATP, dGTP, dCTP, dTTP; ii) = single-stranded DNA; iii) = DNA polymerase III; iv) = Yes
    c. i) = dATP, dGTP, dCTP, dTTP; ii) = single-stranded RNA; iii) = DNA polymerase III; iv) = Yes
    d. i) = dATP, dGTP, dCTP, dTTP; ii) = single-stranded DNA; iii) = DNA polymerase I; iv) = Yes
    e. i) = ATP, GTP, CTP, TTP; ii) = single-stranded DNA; iii) = DNA polymerase III; iv) = Yes

16. In semiconservative replication, what fraction of the double-stranded daughter DNA molecules consists of one original parental strand and one newly synthesized DNA strand after 1, 2, and 3 rounds of replication?
    a. all, a half, an eighth
    b. a half, an eighth, a sixteenth
    c. a half, a quarter, an eighth
    d. all, a half, a quarter
    e. all, a quarter, an eighth

17. During DNA replication, _______ occurs in the 5′ to 3′ direction.
    a. lagging strand DNA synthesis
    b. excision of the RNA primers
    c. proofreading
    d. a and b above
    e. All of the above

18. Choose the option below that provides the correct answers to the following questions about transcription in bacteria:
  i) From what substrates is RNA made?
  ii) On what template?
  iii) With what enzyme?
  iv) Is a primer required?

  a. i) = ATP, GTP, CTP, TTP; ii) = single-stranded RNA; iii) = RNA polymerase; iv) = Yes
  b. i) = ATP, GTP, CTP, UTP; ii) = single-stranded RNA; iii) = RNA polymerase; iv) = No
  c. i) = ATP, GTP, CTP, TTP; ii) = single-stranded DNA; iii) = RNA polymerase; iv) = No
  d. i) = ATP, GTP, CTP, UTP; ii) = single-stranded DNA; iii) = RNA polymerase; iv) = Yes
  e. i) = ATP, GTP, CTP, UTP; ii) = single-stranded DNA; iii) = RNA polymerase; iv) = No

19. Choose the two RNA sequences that could conceivably result from transcription of a double-stranded DNA molecule that has the following sequence in one of its strands.

5′-AGATCCGATGCAT-3′

  a. 5′-UCUAGGCUACGUA-3′, 5′-AUGCAUCGGAUCU-3′
  b. 5′-AGAUCCGAUGCAU-3′, 5′-AUGCAUCGGAUCU-3′
  c. 5′-AGATCCGATGCAT-3′, 5′-AUGCAUCGGAUCU-3′
  d. 5′-AGAUCCGAUGCAU-3′, 5′-UACGUAGCCUAGA-3′
  e. 5′-UCUAGGCUACGUA-3′, 5′-UACGUAGCCUAGA-3′

20. Which one of the following statements about transcription is correct?
  a. RNA polymerase can only add RNA nucleotides to the 5′ end of a growing RNA chain.
  b. The RNA transcript remains hydrogen bonded to the DNA template until transcription of the gene is complete.
  c. Only one of the DNA strands for the entire length of a bacterial or eukaryotic chromosome is used as a template by RNA polymerase.
  d. In bacteria, three different RNA polymerases are needed to transcribe all of the genes.
  e. In some organisms, base-pairing of the RNA transcript to itself is responsible for terminating RNA synthesis.

21. What chemical groups are present at the 5′ end and the 3′ end of a molecule of bacterial messenger RNA (mRNA) <u>immediately</u> after its synthesis?
  a. a 5′ triphosphate group and a 3′ hydroxyl (OH) group
  b. a 5′ monophosphate group and a 3′ hydroxyl (OH) group
  c. a 5′ Shine-Dalgarno sequence and a 3′ poly(A) tail
  d. a 5′ Methyl-guanosine cap structure and a 3′ poly(A) tail
  e. a 5′ start codon and a 3′ stop codon

22. A tryptophan (Trp) amino acid already attached to its tRNA molecule (tryptophan-$tRNA^{Trp}$) was chemically converted to a tyrosine (Tyr). These tyrosine-carrying $tRNA^{Trp}$ molecules (tyrosine-$tRNA^{Trp}$) were then added to a translation system from which all of

the normal tryptophan tRNAs (tryptophan-tRNA^{Trp}) had been removed. What would happen to the resulting protein?

a. Tryptophan would be inserted at every point in the protein chain where tyrosine was meant to be.
b. Tyrosine would be inserted at every point in the protein chain where tryptophan was meant to be.
c. Protein synthesis would stop before completion of the protein chain due to the lack of tryptophan-tRNAs.
d. No tyrosines would be inserted into the protein chain.
e. The tyrosine-tRNA^{Trp} molecules would be unable to base pair to the mRNA template molecule.

23. Which one of the following statements about translation is correct?
a. The large and small subunits of an individual ribosome always stay together.
b. Ribosomes are membrane-bound organelles and therefore are only found in eukaryotic cells.
c. An mRNA molecule binds to a large ribosomal subunit and a tRNA molecule and the small ribosomal subunit then join to form the initiation complex.
d. In eukaryotes, a ribosome and an RNA polymerase molecule can be present on an mRNA at the same time.
e. Multiple different proteins can be synthesized from a single bacterial mRNA.

24. Use the codon table provided below to determine which one of the following nucleotide sequences could not code for the amino acid sequence Leu-Arg-Gln-Ser-Trp.
a. I = 5′-UUA-AGA-CAA-AGC-UGG-3′
b. I = 5′-UUA-AGG-CAA-UCC-UGG-3′
c. I = 5′-CUA-AGA-CAG-AGU-UGG-3′
d. I = 5′-CUA-CGA-CAG-AGA-UGG-3′
e. I = 5′-CUG-CGG-CAG-UCG-UGG-3

	U	C	A	G	
U	Phe Phe Leu Leu	Ser Ser Ser Ser	Tyr Tyr Stop Stop	Cys Cys Stop Trp	U C A G
C	Leu Leu Leu Leu	Pro Pro Pro Pro	His His Gln Gln	Arg Arg Arg Arg	U C A G
A	Ile Ile Ile Met	Thr Thr Thr Thr	Asn Asn Lys Lys	Ser Ser Arg Arg	U C A G
G	Val Val Val Val	Ala Ala Ala Ala	Asp Asp Glu Glu	Gly Gly Gly Gly	U C A G

25. The nucleotides AGC were paired with the nucleotides TCG. You could say with some degree of certainty that this pairing *most likely* did not occur
a. during DNA replication.
b. during transcription.
c. during translation.
d. during transcription or translation.
e. during any of the above.

26. Which of the following processes does not occur in the <u>nucleus</u> of a eukaryotic cell?
a. DNA replication
b. transcription
c. mRNA processing
d. translation
e. mRNA processing and translation

27. Which one of the following mutations would be the most likely to result in the production of an inactive protein?
a. The deletion or insertion of three base pairs close to the start of a gene
b. The deletion or insertion of three base pairs close to the end of a gene
c. The substitution of one base pair
d. The deletion or insertion of one base pair close to the start of a gene
e. The deletion or insertion of one base pair close to the end of a gene

28. Which one of the following types of mutagen is most likely to cause a frameshift mutation?
a. A base analogue
b. An alkylating agent
c. An intercalating agent
d. Ultraviolet radiation
e. All of the above are equally likely to cause a frameshift mutation.

29. When a mutation leads to the replacement of one codon for another that specifies a different amino acid, the resulting mutation is known as a
a. frameshift mutation.
b. missense mutation.
c. nonsense mutation.
d. samesense mutation.
e. deletion mutation.

Answers

1. **F** 2. **T** 3. **T** 4. **F** 5. **T** 6. **e** 7. **b** 8. **c** 9. **e** 10. **b** 11. **e** 12. **a**
13. **a** 14. **e** 15. **b** 16. **d** 17. **d** 18. **e** 19. **b** 20. **e** 21. **a** 22. **b** 23. **e**
24. **d** 25. **c** 26. **d** 27. **d** 28. **c** 29. **b**

CHAPTER 5

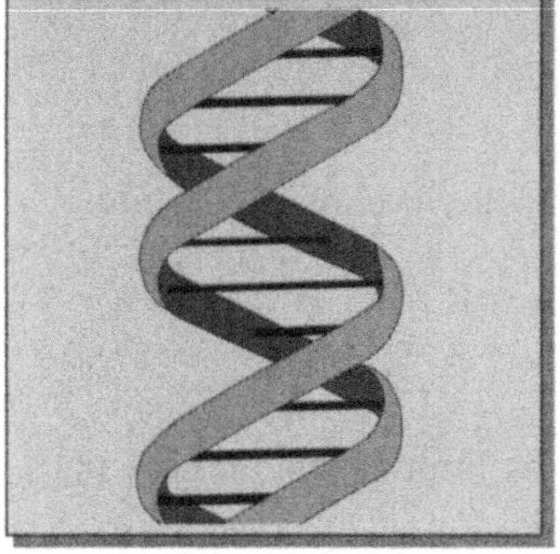

Regulation of Gene Expression in Bacteria

As discussed in previous chapters, the processes of transcription and translation consume large amounts of energy so that expressing all genes in a cell all the time would be extremely energy intensive. However, it does not make sense for an organism to use valuable ATP to make enzymes if they are not needed. For example, many bacterial enzymes are needed to metabolize foodstuffs present in the environment, and it does not make sense for a bacterium to make these enzymes when those foods are not present. In fact, many genes are not expressed all the time; instead, they are switched on and off as their products are needed by the cell. Genes that are turned on and off according to the needs of the cell are said to be **regulated**. Some genes are not regulated but are expressed all the time because their products are always needed by the cell. Such continually expressed genes are said to be **constitutive** and include, for example, those genes whose products are involved in protein synthesis and cellular respiration.

ESSENTIAL BACKGROUND

- **Transcription and translation of bacterial genes**
- **Basic characteristics of bacterial cells**

TOPIC 1: REGULATED VERSUS CONSTITUTIVE GENES

KEY POINTS

✓ *Why is the expression of many bacterial genes regulated?*

✓ *What is the primary role of gene regulation in bacteria?*

Most bacteria are free-living unicellular organisms that grow and divide indefinitely as long as environmental conditions are suitable. Thus, in bacteria, the main role of gene regulation is to enable individual bacteria to adjust their metabolism to achieve maximum growth rate in a particular environment.

Obviously, gene regulation can take the form of any mechanism that affects the production of mRNA or protein or affects the activity of proteins. Mechanisms that regulate each step from DNA to protein activity exist in bacteria; however, most gene regulation occurs at the level of transcription, because the earlier gene expression is stopped, the more energy a cell saves itself.

Topic Test 1: Regulated Versus Constitutive Genes

True/False

1. Because bacterial genomes generally contain far fewer genes than eukaryotic genomes, all genes in bacteria are expressed all the time.
2. Regulation of gene expression in bacteria occurs most often at the level of translation.

Multiple Choice

3. What is the primary role of gene regulation in bacteria?
   a. To conserve the molecules of RNA polymerase in the cell by limiting the number of genes being transcribed at any one time.
   b. To differentiate cells for specific functions in the organism.
   c. To adjust rates of protein synthesis to allow the cell to achieve a maximum growth rate.
   d. To maximize the synthesis of proteins from specific genes when their gene products are required by the cell.
   e. To adjust the level of ATP produced in the cell to meet the demands in a particular environment.

Short Answer

4. Genes that are turned on and off according to the needs of the cell are said to be ___________.
5. What is meant by constitutive expression?

Topic Test 1: Answers

1. **False.** Despite the fact that bacterial genomes contain far fewer genes than eukaryotic genomes, it would still be an extreme waste of energy and other valuable resources for a bacterial cell to make gene products that it does not need. It is also particularly important that bacteria can adjust their metabolic levels to adapt to their external environment, which typically changes more frequently than that surrounding a eukaryotic cell. In some cases, continuing to synthesize a gene product when it is no longer needed by the cell can actually be detrimental to the growth of the bacterium.
2. **False.** Because more energy is saved the earlier gene expression is stopped, most regulation occurs at the level of transcription, not translation.
3. **c.** Regulation occurs to maximize synthesis of gene products that are required at any one particular time and to minimize synthesis of gene products that are not required. There is no "defined" number of genes that are expressed at all times, because different environments can result in drastically different metabolic requirements. a is not correct because the genes for RNA polymerase are constitutively (constantly) expressed so RNA polymerase is not limiting. b is not a correct statement because bacteria are unicellular organisms and therefore do not produce different cell types specialized for different functions. Likewise, d is incorrect because it is only half the answer. It is just as important to prevent the expression of genes whose products are not required at the time as it is to

maximize expression of genes whose products are required. e is incorrect because regulation affects gene expression levels, not energy availability.

4. **Regulated.** For example, many genes whose products are required to break down foodstuffs are only expressed when those foods become available.

5. Constitutive expression refers to the continual transcription of certain genes whose products are always needed by the cell. Constitutively expressed genes include, for example, the genes whose products are involved in cellular respiration and protein synthesis.

TOPIC 2: TRANSCRIPTIONAL REGULATION AND OPERONS

KEY POINTS

✓ *What is an operon?*

✓ *What keeps the genes for lactose metabolism (the* lac *genes) "turned off" in the absence of lactose?*

✓ *What turns the* lac *genes on in the presence of lactose?*

✓ *How does glucose prevent efficient expression of the* lac *genes if both glucose and lactose are present in the growth medium?*

✓ *Does gene regulation always result in genes being "turned* on*"?*

If several enzymes are involved in the same pathway, it usually follows that either all or none of the enzymes are made. The simultaneous production of functionally related enzymes is coordinated by transcribing all the related genes into a single **polycistronic mRNA**. Such coordinately regulated units consisting of closely linked structural genes and the stretches of DNA that control their transcription are known as **operons**, which are found only in bacteria: eukaryotes do *not* contain operons.

Transcriptional regulation falls into one of two major categories: **positive regulation** or **negative regulation**. In negative regulation, the gene is "turned off" (or expressed at very low levels) by an inhibitor that binds to the DNA and prevents RNA polymerase from transcribing the gene. The inhibitor has to be removed for transcription to occur (**Figure 5.1**). In positive regulation, the gene is turned off because, in this case, transcription requires an activator molecule that is needed to enhance the binding of RNA polymerase, thus turning on the gene (Figure 5.1). The lactose operon described below is a good model for explaining transcriptional regulation because it is both negatively and positively regulated.

LACTOSE OPERON OF *ESCHERICHIA COLI*

In *E. coli* two proteins are known to be needed for the metabolism of lactose. **Lactose permease** (product of the *lacY* gene) is needed to transport lactose into the cell, and the enzyme **β-galactosidase** (product of the *lacZ* gene) is needed to cleave lactose into glucose and galactose. The genes for these two proteins, plus the *lacA* gene that codes for transacetylase, a protein of as yet unknown function in lactose metabolism, are under the control of a single promoter and are transcribed into a single polycistronic mRNA (**Figure 5.2**).

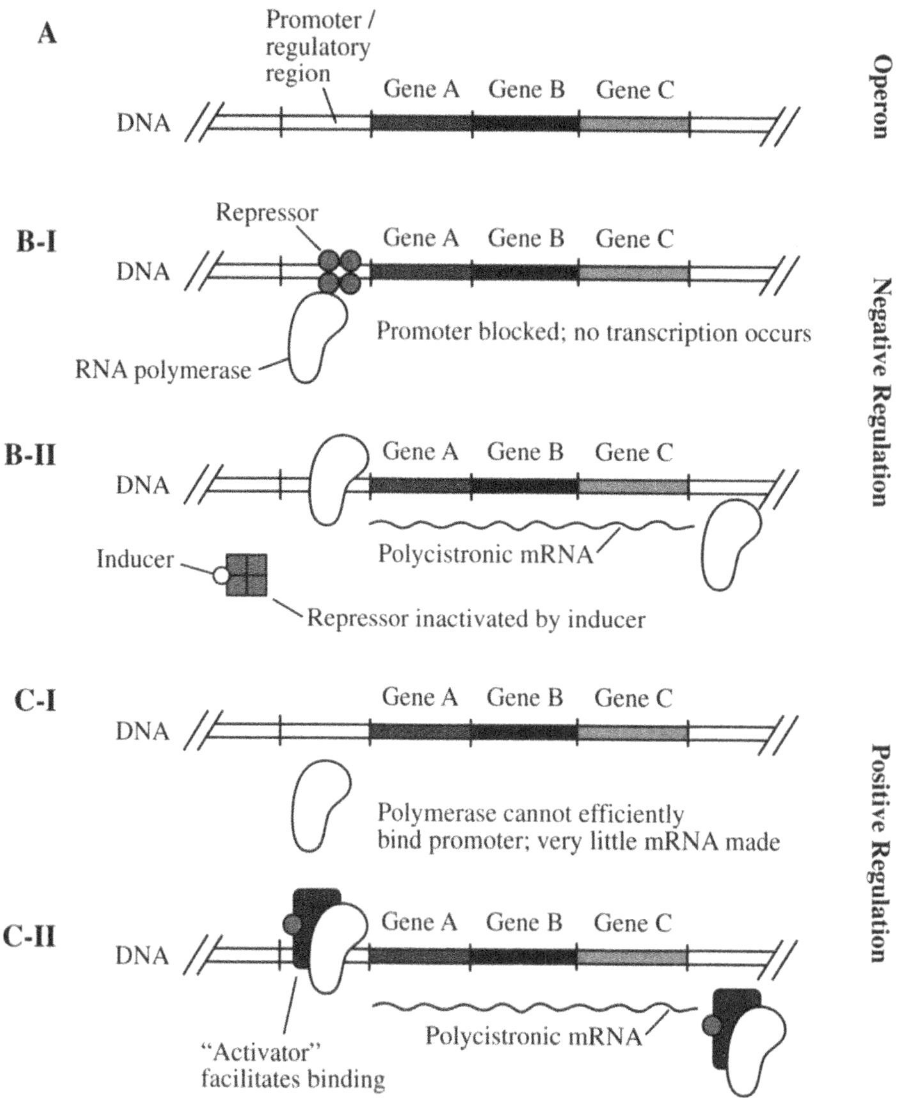

Figure 5.1 Comparison of positive and negative regulation. A typical operon is shown in A, with three structural genes preceded by a regulatory region including the single promoter site for the operon. B-I and B-II demonstrate negative regulation. In B-I, a repressor protein is bound to the regulatory region, preventing RNA polymerase from binding to the promoter and blocking expression of the operon. Binding of an "inducer" molecule to the repressor causes it to release the operator. Therefore, in the presence of an inducer molecule, RNA polymerase is free to bind the promoter, and synthesis of the polycistronic mRNA occurs, as illustrated in B-II. Positive regulation is illustrated in C-I and C-II. RNA polymerase alone cannot bind the promoter efficiently and little or no mRNA is made. However, in the presence of an "activator" molecule, polymerase binding is very efficient, and a large amount of mRNA is synthesized.

In the absence of lactose, the levels of all three of these proteins are extremely low. *E. coli* does not waste valuable resources making these proteins when their substrate, lactose, is not present. However, if the environment in which the *E. coli* is growing (e.g., the human intestine) suddenly changes such that lactose becomes the predominant sugar, the synthesis of all three of these proteins increases dramatically. Thus, in the absence of lactose, there are only two to three molecules of β-galactosidase in an *E. coli* cell compared with 3,000 to 6,000 molecules of

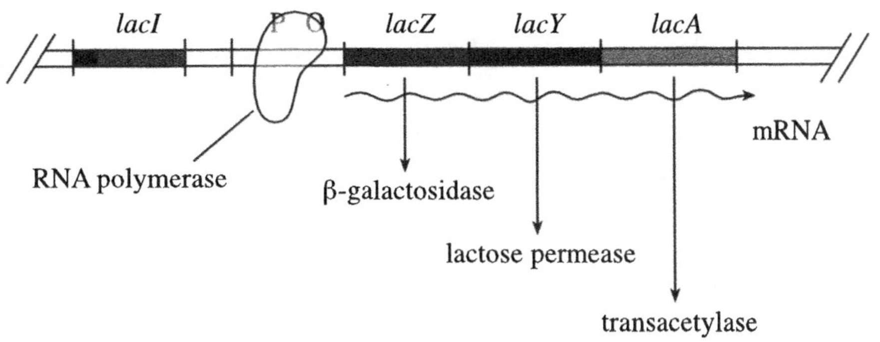

Figure 5.2 Organization of the lactose (*lac*) operon in *E. coli*. All three structural genes of the *lac* operon (*lacZ*, *lacY*, and *lacA*) are transcribed into a single polycistronic mRNA but are translated into three individual proteins as described in the text. The structural genes are preceded by a regulatory region that consists of the promoter site (P) and an operator region (O), which is the binding site for the *lac* repressor protein, the product of the *lacI* gene. *lacI* lies a short distance away from the *lac* operon genes and is controlled by its own promoter sequence (the black region in the area labeled *lacI*).

β-galactosidase in the presence of lactose. What keeps the genes for lactose metabolism (the *lac* genes) turned off in the absence of lactose? What turns the *lac* genes on in the presence of lactose?

In addition to the *lac* operon genes mentioned above, the system responsible for metabolizing lactose involves a regulatory gene called the ***lacI* gene** (Figure 5.2). The *lacI* gene, which lies close to the *lac* operon, is expressed all the time at a slow rate and its product, the ***lacI* repressor**, binds to a specific site called the **operator** that lies within the promoter for the *lac* operon.

NEGATIVE REGULATION OF THE LACTOSE OPERON

In the absence of lactose, the *lacI* repressor produced by the *lacI* gene binds to the operator and in so doing blocks the binding of RNA polymerase to the *lac* promoter (**Figure 5.3**). Therefore, when the repressor is bound to the operator, it prevents expression of the *lac* genes, so the *lac* operon is said to be negatively regulated by the repressor. When lactose becomes available, **allolactose**, a derivative of lactose that is automatically produced in the cell when lactose is present, binds to the repressor protein and alters its shape (allosteric modification) in such a way that the repressor can no longer bind to the operator. Thus, in the presence of lactose, the operator is unoccupied so RNA polymerase can bind to the *lac* promoter and start transcribing the *lac* genes. Compounds that stimulate transcription (as does allolactose/lactose in this example) are called **inducers** and the resulting proteins that are produced are said to be **inducible**.

Note that for the operator to exert its effects it has to be adjacent to the *lacZ*, *lacY*, and *lacA* genes. The *lacI* gene, on the other hand, does not need to be close to the operon to exert its effects because the repressor is a soluble protein and is able to diffuse throughout the cell. Most of our knowledge of the regulation of the *lac* operon comes from studying mutants that are unable to metabolize lactose or that show constitutive (unregulated) expression of the *lac* genes. To test your understanding of how different mutations affect expression of the *lac* operon, see the Demonstration Problem at the end of this chapter.

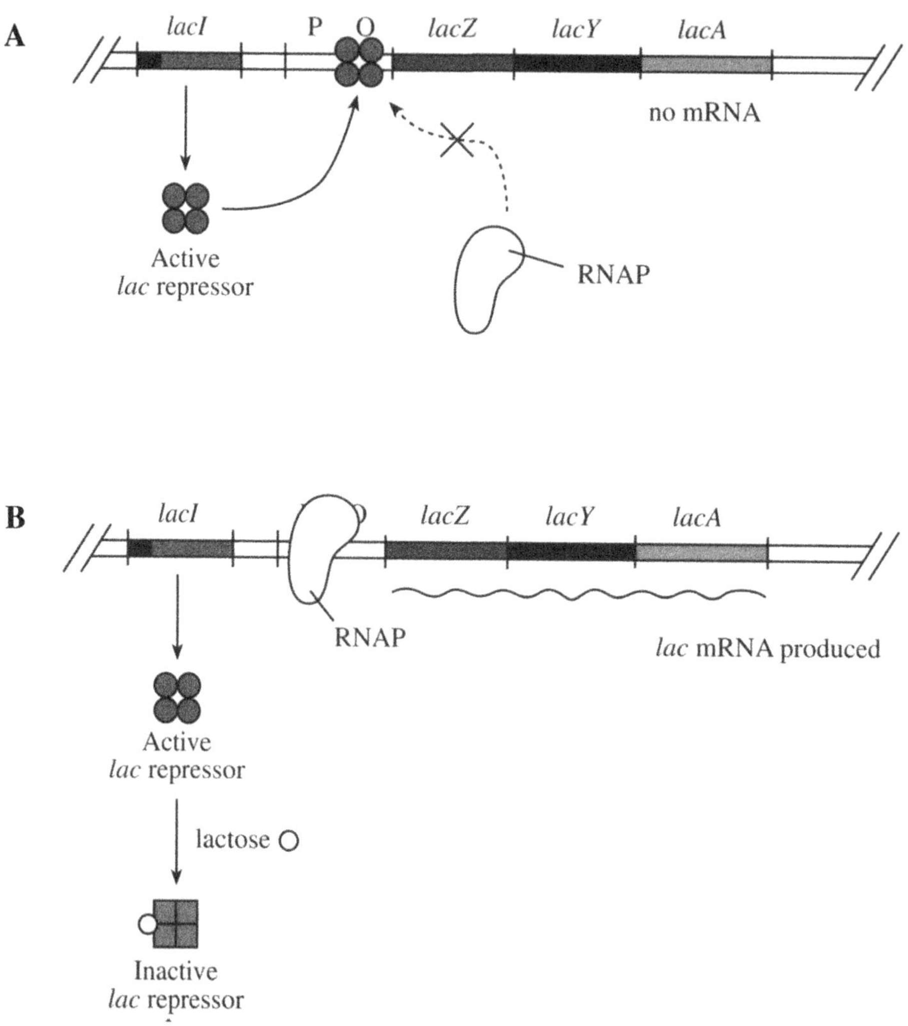

Figure 5.3 Negative regulation of the lactose operon. A. In the absence of lactose, the *lac* repressor (the product of the *lacI* gene) binds to the operator region (O) of the *lac* operon. When the repressor is bound to the operator, it prevents RNA polymerase (RNAP) from accessing the *lac* promoter (P), thereby blocking transcription of the *lac* genes. B. When lactose is present, it binds to, and changes the conformation of, the *lac* repressor. The inactivated repressor protein is now incapable of binding to the operator region (O), allowing RNA polymerase (RNAP) access to the *lac* promoter (P). Transcription of the *lac* genes can now occur.

POSITIVE REGULATION OF THE LACTOSE OPERON: THE EFFECT OF GLUCOSE

Further studies of lactose metabolism in *E. coli* revealed that very inefficient transcription of the *lac* genes occurs if *both* glucose and lactose are present in the growth medium. Based on the previous discussion of the effects of lactose on transcription of the *lac* operon, it was immediately obvious that another regulatory molecule must be needed for efficient transcription of the *lac* genes and furthermore that this regulatory molecule must be itself regulated by the concentration of glucose.

E. coli prefers to metabolize glucose over all other sugars. For this reason, a mechanism has evolved to prevent the efficient expression of genes whose products are needed for the metabolism of sugars other than glucose when glucose is also present in the environment.

Table 5.1 Effect of Glucose on Levels of cAMP and *lac* Gene Expression

Carbon Source	*lac* Gene Expression*	Concentration of cAMP
Glycerol (no lactose, no glucose)	Off	High
Lactose + glucose	Off	Low
Lactose (no glucose)	On	High
Glucose (no lactose)	Off	Low

* For convenience, the term "off" is used but few genes, if any, are ever completely turned off. Instead, in the off state, a basal level of gene expression almost always remains, in the case comprising two or three molecules of β-galactosidase per cell generation.

CYCLIC AMP–CYCLIC AMP RECEPTOR PROTEIN COMPLEX

E. coli (and many other bacterial species) contain a protein called the **cyclic AMP-receptor protein** or **CRP** for short that is encoded by the *crp* gene. CRP binds to **cyclic AMP** (**cAMP**) and the resulting cAMP–CRP complex binds to the *lac* promoter close to the binding site for RNA polymerase. In this position, the cAMP–CRP complex greatly *enhances* the binding of RNA polymerase and hence transcription of the *lac* genes.

It is cAMP that links transcription of the *lac* operon to the concentration of glucose. High concentrations of glucose inhibit adenylyl cyclase, the enzyme responsible for the synthesis of cAMP. Therefore, when the level of glucose in the cell is high, the cAMP concentration is low. Conversely, when the supply of glucose has been exhausted, requiring that *E. coli* find an alternative energy source (e.g., lactose), the concentration of cAMP rises, so the cAMP–CRP complex can form, resulting in efficient transcription of the *lac* genes (**Figure 5.4** and **Table 5.1**). The cAMP–CRP mechanism of transcriptional regulation is an example of **positive regulation** because binding of the cAMP–CRP complex enhances transcription (compare with negative regulation of the *lac* operon where binding of the repressor protein prevents transcription).

cAMP–CRP COMPLEX IS A GLOBAL REGULATOR

The cAMP–CRP complex is said to be a global regulator because it is responsible for enhancing the binding of RNA polymerase to the promoters of many genes responsible for the metabolism of sugars.

TRYPTOPHAN (*trp*) OPERON

It is very economical for *E. coli* to be able to switch on the *lac* genes only when lactose is present and glucose is absent from the environment. Equally valuable to a bacterium is the ability to switch off the synthesis of enzymes if their products accumulate or become available in the environment. For example, the amino acid tryptophan is an essential constituent of proteins, and *E. coli* (and other bacteria) needs an ample supply of tryptophan for cell growth. The *trp* operon consists of a promoter, an operator, and five structural genes (*trpE*, *trpD*, *trpC*, *trpB*, and *trpA*) that encode the enzymes necessary for tryptophan biosynthesis. Regulation of the *trp* operon is exerted by the *trp* repressor, the product of the *trpR* gene that is located elsewhere on the chro-

Table 5.2 Inducible Versus Repressible Operons	
INDUCIBLE	**REPRESSIBLE**
Operon off—genes need to be turned on	Operon on—genes need to be turned off
Repressor alone can bind to operator	Repressor requires a corepressor to bind to operator
Operon turned on by substrate for operon enzymes	Operon turned off by external availability of product of operon enzymes
Example, most enzymes that break down external metabolites	Example, most enzymes involved in the synthesis of essential amino acids and vitamins

mosome. As with the *lac* operon, the *trp* repressor binds to the operator in the *trp* promoter region and prevents expression of the *trp* genes. However, in contrast to the *lac* operon which is normally off in the absence of lactose, in the absence of tryptophan, the *trp* operon is on as, under these conditions, the bacterium needs to generate its own supply of tryptophan. The reason the *trp* genes are expressed in the absence of tryptophan is because the *trp* repressor by itself cannot bind to the operator but needs to be activated by the binding of a **corepressor**. In the *trp* operon, the end product, tryptophan, itself acts as the corepressor, binding to the *trp* repressor to achieve repression. Thus, only when tryptophan is present in the environment can an active repressor form and inhibit transcription of the *trp* genes. When the external supply of tryptophan is depleted, the co-repressor dissociates from the repressor, the operator is vacated, and transcription of the *trp* genes begins again.

Because the synthesis of tryptophan is turned off in response to the external availability of tryptophan, synthesis of tryptophan is said to be **repressible** (compare with the *lac* genes, which are **inducible**—turned on by the presence of lactose; **Table 5.2**). The expression of many genes involved in the synthesis of essential amino acids and vitamins is regulated in this fashion.

Topic Test 2: Transcriptional Regulation and Operons

True/False

1. It is necessary for the operator to be within or immediately adjacent to the promoter that it regulates.
2. It is necessary for the repressor gene to be adjacent to the operator that it regulates.
3. The function of lactose permease is to transport lactose into the cell.

Multiple Choice

4. Which one of the following statements about regulation of the *lac* operon is correct?
   a. The cAMP–CRP complex binds to the *lac* promoter and blocks the binding of RNA polymerase.
   b. The *lacI* repressor negatively regulates expression of the *lacI*, *lacZ*, *lacY*, and *lacA* genes.
   c. The *lacI* repressor binds to the operator within the *lac* promoter and enhances the binding of RNA polymerase.
   d. In the absence of glucose and the presence of lactose, RNA polymerase has a strong affinity for the *lac* promoter.

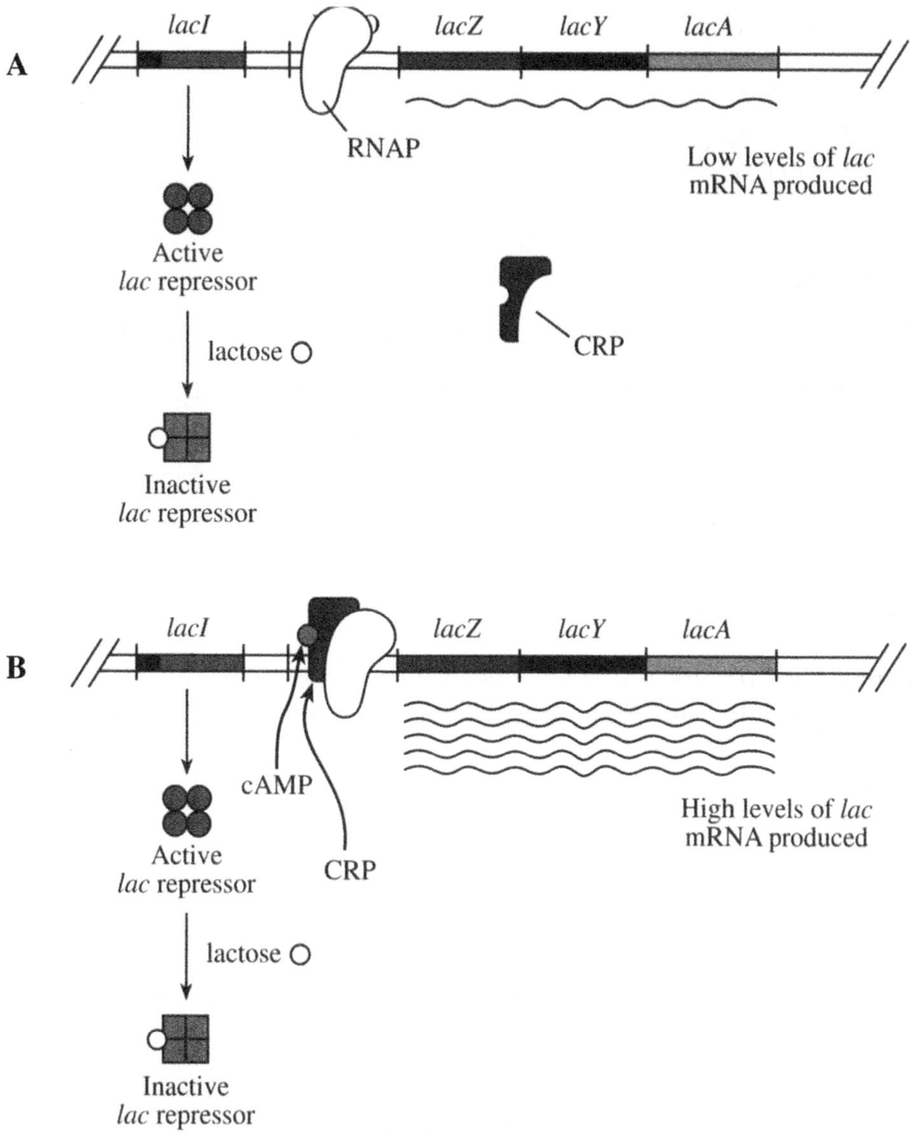

Figure 5.4 Positive regulation of the lactose operon. A. *E. coli* growing in medium containing both glucose and lactose. Lactose binds to the *lac* repressor, inactivating the repressor and allowing RNA polymerase access to the promoter. However, in the presence of glucose, the levels of cAMP are low, preventing formation of the cAMP–CRP complex. Without the assistance of this complex, RNA polymerase binds only weakly to the promoter, resulting in low levels of *lac* mRNA. B. *E. coli* growing in medium containing lactose but no glucose. As before, the *lac* repressor is inactivated, but the absence of glucose means high levels of cAMP, which complexes with CRP. The cAMP–CRP complex *positively regulates* expression of the *lac* operon by enhancing the affinity of RNA polymerase for the *lac* promoter, yielding high levels of *lac* mRNA.

e. In the presence of glucose and lactose, the *lacI* repressor is bound to the operator within the *lac* promoter.

5. *E. coli* cells with which one of the following genotypes would produce active β-galactosidase only in the *presence* of lactose? Note, the mutation in the *lac* operator sequence (denoted O^C) prevents the *lacI* repressor from binding to the operator but has no effect on binding of RNA polymerase to the *lac* promoter.

a. *lacI*$^+$ *lacO*$^+$ *lacZ*$^+$
b. *lacI*$^-$ *lacO*$^+$ *lacZ*$^+$
c. *lacI*$^-$ *lacO*^{C} *lacZ*$^+$
d. *lacI*$^+$ *lacO*^{C} *lacZ*$^+$
e. *lacI*$^+$ *lacO*$^+$ *lacZ*$^-$

Short Answer

6. Many functionally related bacterial genes are organized into clusters called ___________.
7. Each of these clusters is transcribed from a single ___________.
8. A ___________ protein binds to an operator sequence within the regulatory region for a gene cluster, as described in question 7, and blocks transcription. This is an example of what type of regulation?

Topic Test 2: Answers

1. **True.** The operator is a DNA sequence that is bound by a regulatory protein—it is not transcribed into mRNA and translated into a protein product. Therefore, for the regulatory protein that binds to the operator to affect RNA polymerase binding to the promoter, the operator must be within or immediately adjacent to the promoter that it regulates.
2. **False.** The repressor gene is transcribed into mRNA and translated into the repressor protein. Once synthesized, this repressor protein diffuses freely within the cell and is capable of binding to any appropriate operator sequence. It is therefore not necessary for the repressor gene to be adjacent to the operator site that the repressor protein will eventually bind to.
3. **True.** Lactose cannot freely diffuse across the cell wall but has to be actively transported into the bacterial cell by lactose permease.
4. **d.** The cAMP–CRP complex enhances binding of RNA polymerase rather than blocks it, as suggested in a. In b, the *lacI* repressor does negatively regulate expression of *lacZ*, *lacY*, and *lacA* (the genes in the *lac* operon), but it does not regulate expression of *lacI*. Moreover, because the *lacI* repressor negatively regulates expression of the *lac* operon genes, it inhibits binding of RNA polymerase rather than enhancing it, as stated in c. Finally, e is incorrect because the presence or absence of glucose does not change the fact that lactose itself binds the *lacI* repressor and removes it from the operator. In the presence of lactose, the repressor will not be bound to the operator so RNA polymerase can bind to the *lac* promoter. If glucose is also absent (conditions described in d), the cAMP–CRP complex will be available to enhance RNA polymerase binding.
5. **a.** Under wild-type conditions in the absence of lactose, the product of the *lacI* gene, the *lacI* repressor, binds to the *lac* operator and inhibits expression of the *lac* genes. In the presence of lactose, the repressor is inactivated, regulation is lifted, and expression of β-galactosidase occurs. This situation is best described by a. b has a mutation in the *lacI* gene, meaning either that no repressor protein will be produced or that the mutant repressor protein produced will be unable to bind to the operator. This eliminates the

negative regulation mechanism, causing β-galactosidase to be expressed even in the absence of lactose. Likewise, an O^c mutation would prevent binding of the repressor protein, also leading to constitutive (constant) production of β-galactosidase; this is the case in d. c contains both of these mutations (*lacI*$^-$ and *lacO*^{C}) and will exhibit constitutive expression of β-galactosidase. Regulation is intact in e, but there is a mutation in the gene for β-galactosidase, so active β-galactosidase will not be produced.

6. **Operons.** Many functionally related bacterial genes are grouped together under the control of a single promoter so that their expression can be coordinately regulated. Such a coordinately regulated unit is called an operon and is transcribed into a single polycistronic mRNA.
7. **Promoter.** Every transcriptional unit, whether comprised of one gene or several (as in an operon), must have a promoter region dictating the binding site for RNA polymerase.
8. **Repressor.** Regulation that blocks or impairs expression of genes is called negative regulation. Regulatory proteins that enhance expression would exhibit positive regulation.

TOPIC 3: TRANSLATIONAL AND POSTTRANSLATIONAL REGULATION OF BACTERIAL GENES

KEY POINTS

✓ *How can different proteins translated from the same polycistronic mRNA be produced in different amounts?*

✓ *Why do bacterial mRNAs have such short half-lives?*

✓ *How does the bacterial cell inhibit the activity of enzymes that are already present in the cell if that enzyme activity is no longer needed?*

TRANSLATIONAL REGULATION

The amount of protein made from a particular mRNA is determined by the frequency with which a ribosome binds the ribosome binding site (RBS) that precedes the initiator codon. For example, the ratio of the number of copies of β-galactosidase, permease, and transacetylase made from the *lac* mRNA is approximately 10:5:2, and this variation reflects differences in efficiency of ribosome binding to the individual RBS sequences. After a ribosome has translated the *lacZ* gene, it usually dissociates from the *lac* mRNA. The frequency at which the next gene, *lacY*, is translated is determined by the efficiency with which a ribosome binds to the RBS sequence preceding the *lacY* gene and the same applies for the translation efficiency of the *lacA* gene.

In addition to differences between the individual RBS sequences (the closer the RBS sequence is to the consensus sequence, the higher the frequency of ribosome binding), the efficiency of ribosome binding is also known to be affected by the formation of secondary structures in this region.

MESSENGER RNA (mRNA) STABILITY

The amounts of the three Lac proteins also depend on the availability of their corresponding mRNA segments. Bacterial mRNAs are unstable, with half-lives on the order of a few minutes. This rapid turnover of the mRNA population enables bacteria to respond rapidly to radical changes in their environment. The fact that degradation of *lac* mRNA is initiated from its 3′ terminus also explains the reason for the 10 : 5 : 2 ratio as, at any given time, there will be more copies of the *lacZ* part of the mRNA than of the *lacY* part of the mRNA and more copies of the *lacY* mRNA than the *lacA* mRNA.

ALLOSTERIC REGULATION AND FEEDBACK INHIBITION

The transcriptional regulation mechanisms discussed in terms of the regulation of the *lac* and *trp* operons allow the cell to adapt to changes in the environment by changing the concentrations of key enzymes. This form of regulation results in great savings in energy but is slow to take effect. For example, when a large amount of tryptophan is available in the environment, expression of the *trp* genes is repressed because tryptophan synthesis is unnecessary. However, if the cell contains a large amount of the tryptophan enzymes before repression occurs, they could theoretically continue to synthesize tryptophan even though the cell no longer needs to make its own. To prevent the cell from wasting valuable resources to make a product that is readily available in the environment, mechanisms have evolved that allow for rapid adjustment to changes in the environment by adjusting the **activity** of enzymes already present in the cell. Inhibition of enzyme activity is achieved by small regulatory molecules that bind to a site different from the active site and change the shape of the enzyme in such a way that it is no longer able to recognize its substrate. Enzymes (proteins) whose shape is changed by binding of a small molecule to a site different from the active site are called **allosteric enzymes**. The site on an allosteric enzyme that the regulatory molecule binds to is called the allosteric site, and the resulting change to the enzyme is referred to as **allosteric modification**.

In a biochemical pathway, the end product of the pathway is typically an allosteric inhibitor of the enzyme that catalyzes the **first committed step** in that pathway, that is, the first step in the pathway that leads to the synthesis of only that end product and no other. This is the principle of **feedback inhibition** or **end product inhibition**. Allosteric inhibition is usually completely reversible—when the concentration of end product falls to a low level, the end product releases the enzyme, which becomes active again. An example of a pathway that is regulated by feedback inhibition is glycolysis (**Figure 5.5**). In this case, the allosteric inhibitor is one of the end products, ATP, and the enzyme that is allosterically modified by the ATP is phosphofructokinase, which catalyzes the first committed step in glycolysis (the first step that leads to the synthesis of pyruvate and no other product).

Topic Test 3: Translational and Posttranslational Regulation of Bacterial Genes

True/False

1. All proteins translated from a single polycistronic mRNA are made in the same quantity.
2. Bacterial mRNA molecules are relatively unstable and are often degraded after only a few minutes.

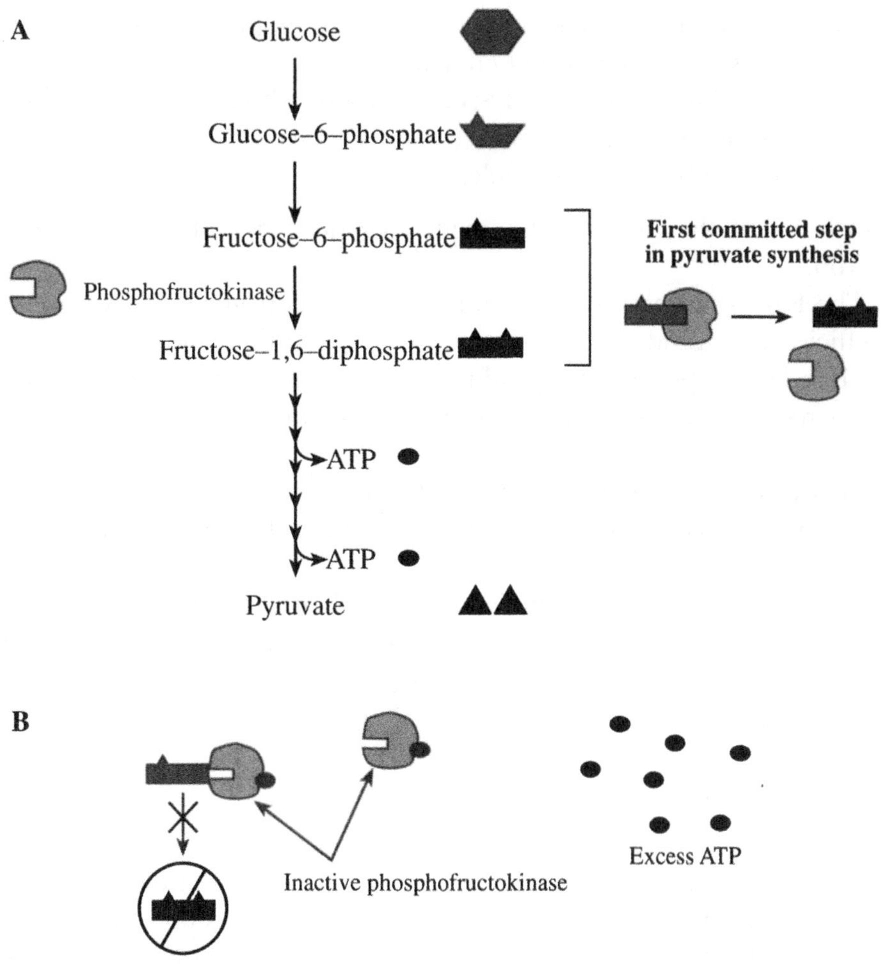

Figure 5.5 Feedback inhibition. (A) Glycolysis and the first committed step in the production of pyruvate. Glucose is sequentially broken down to pyruvate, with each step catalyzed by a different enzyme. The first step unique to pyruvate production is the conversion of fructose-6-phosphate to fructose-1,6-diphosphate by the enzyme phosphofructokinase. Production of ATP occurs late in the pathway as shown in the figure. (B) Feedback inhibition by ATP. When the level of ATP in the cell exceeds demand, excess ATP binds to an allosteric site on phosphofructokinase, causing a conformational change that inhibits the enzyme. Production of pyruvate (and therefore ATP) slows down so that the level of ATP in the cell drops.

Short Answer

3. What generally acts as the inhibitor of the first committed step in a biochemical pathway?
4. A polycistronic mRNA is synthesized, which carries five gene sequences, A, B, C, D, and E in order from 5′ to 3′ is synthesized. Give several reasons why the cell might have less of protein E than protein A.

Multiple Choice

5. Which one of the following statements about feedback inhibition is incorrect?
   a. Feedback inhibition typically occurs at the first committed step in a pathway.
   b. Regulation by feedback inhibition is more energy efficient than regulation at the transcriptional level.

c. Feedback inhibition usually involves binding to an allosteric site that is different from the active site of the affected enzyme.
d. Feedback inhibition of glycolysis occurs through inactivation of the enzyme phosphofructokinase.
e. Feedback inhibition is reversible.

6. Which one of the following statements best describes a feature of regulation of the *lac* operon?
   a. The RBS sequence at the 5′ end of the polycistronic mRNA is used more or less efficiently to produce differing amounts of the three *lac* proteins.
   b. More lactose permease is produced than β-galactosidase because lactose permease is permanently altered as it transports lactose into the cell and has to be continually replaced.
   c. *lac* mRNA is degraded from the 5′ to the 3′ end.
   d. The rapid turnover of *lac* mRNA allows the cell to respond quickly to changes in lactose availability.
   e. Translation of the *lacA* gene is not as efficient as translation of the *lacZ* and *lacY* genes because the promoter sequence for the *lacA* gene is not as close to the consensus promoter sequence as it is for the *lacZ* and *lacY* genes.

Topic Test 3: Answers

1. **False.** The different proteins synthesized from a polycistronic mRNA may have functions that require a small amount of one protein and a large amount of another. For this reason, each gene sequence in a polycistronic mRNA is preceded by its own RBS and terminated by its own stop codon. The efficiency of binding of the ribosome to the RBS determines the amount of protein translated from each gene sequence in the mRNA.
2. **True.** This rapid turnover enables bacteria to respond rapidly to radical changes in their environment.
3. The end product of the pathway generally acts as the inhibitor of the first committed step, a process referred to as feedback inhibition.
4. One reason there may be more protein A than protein E is because the RBS sequence preceding gene A is closer to the consensus RBS sequence than that of gene E. A second reason is that gene E lies at the 3′ end of the mRNA, and, since mRNAs are normally degraded starting from the 3′ terminus, at any given time, there will be more copies of the gene A part of the mRNA than of the gene E part of the mRNA, allowing more time for translation of protein A. Finally, synthesis of protein E may be inhibited by the formation of secondary structures close to the 3′ end of the mRNA that block ribosome binding.
5. **b.** Regulation by feedback inhibition takes effect faster than regulation at the level of transcription, but because it occurs later in the path from DNA to an active protein, it is less energy efficient. All other options correctly describe feedback inhibition.
6. **d.** Each gene on a polycistronic mRNA has its own RBS sequence, which is the binding site for ribosomes. This eliminates a as a possibility, because there is not only one RBS sequence. It also eliminates e, because each gene sequence on a polycistronic mRNA is preceded by its own RBS sequence not by its own promoter sequence. b is incorrect because more β-galactosidase is produced than lactose permease or transacetylase, and

enzymes are not permanently altered by the reactions they catalyze so they can be used over and over. c is not correct because *lac* mRNA is actually degraded from the 3′ end. Only d correctly describes a feature of *lac* operon regulation.

IN THE CLINIC

For expression of a eukaryotic gene in a bacterial host cell, it is necessary to use promoter and termination sequences that are compatible with bacterial RNA polymerase. Because many biotechnology applications require the production of large amounts of protein, specially engineered promoters that are capable of supporting transcription at a high level are often used. For many cloned genes, further manipulations are necessary to ensure stability of their proteins and, if appropriate, their secretion. Cloned gene proteins are much easier to purify if they are secreted or exported out of the host cell. Normally, an amino acid sequence called a signal peptide (or leader peptide) located at the N-terminal end of a protein facilitates its export, and many cloned gene proteins are engineered for secretion by adding the DNA sequence encoding a signal peptide to the cloned gene. However, in many cases, the cloned gene protein is degraded in the bacterial cell before it can be secreted. The degradation of foreign proteins by proteolytic enzymes of the host cell can be avoided by using specially engineered bacterial strains deficient in the production of some of their proteolytic enzymes. Another strategy involves the addition of one or more amino acids to the N-terminus of the protein; often the presence of a single extra amino acid at the N-terminal end is sufficient to ensure stability. Therefore, knowledge of the mechanisms of bacterial gene regulation has permitted the development of bacterial strains that are engineered for high-level production of cloned gene proteins.

DEMONSTRATION PROBLEM

E. coli cells that carry more than one copy of some of their chromosomal genes are called **merodiploids**. The second gene copy is typically introduced into the cell on a plasmid.

1. Determine whether the following statement is **correct** or **incorrect**.

A cell with the following merodiploid genotype would efficiently synthesize functional β-galactosidase only in the presence of lactose.

$$lacI^- \ lacO^+ \ lacP^+ \ lacZ^+ \ Y^+ \ A^+ / lacI^+ \ lacO^+ \ lacP^+ \ lacZ^+ \ Y^+ \ A^+$$

To answer the broader question of whether functional β-galactosidase would be made in this cell, first ask a series of smaller questions. For example,

Is there active *lacI* repressor protein in the cell?

The repressor protein is the product of the *lacI* gene, and it binds to the operator sequence (*lacO*). The repressor protein is diffusible, which means that as long as there is at least one functional copy of the *lacI* gene in a cell, all wild-type lac operator sequences in that cell will be blocked (remembering that many copies of the repressor protein are made from a single *lacI* gene).

If each copy of *lacI* in the cell was *lacI*$^-$, *lacI* repressor protein would not be made so there would be no regulatory mechanism in place to prevent expression of the *lac* structural genes in the absence of lactose.

However, in the example given above, although one copy of the *lac* operon (on the left) has a mutated *lacI* gene and therefore cannot produce functional *lacI* repressor protein, the second copy of the *lac* operon (on the right) has a wild-type *lacI* gene, so there will be *lacI* repressor protein in this cell. If *lacI* repressor protein is present in a cell, the next question that should be asked is:

Can the *lacI* repressor protein bind to both operator sites in the cell?

To answer this question, examine *lacO* in each copy of the *lac* operon. *lacO*$^{+}$ indicates a wild-type operator sequence that can bind the *lacI* repressor protein and hence will be blocked in the absence of the inducer, allolactose (a derivative of lactose). *lacO*^{C} indicates a mutation (called a *c*onstitutive mutation) in the operator site that prevents the *lacI* repressor protein from binding. If any copy of *lacO* in a cell is *lacO*^{C}, expression of the *lac* genes **on that copy of the *lac* operon only** will be constitutive, or continuous in the absence of lactose, as long as RNA polymerase can bind to the promoter (i.e., as long as the promoter itself is not mutated).

Our merodiploid scenario consists of two *lac* operator sequences, both of which are wild type (*lacO*$^{+}$). So the *lacI* repressor protein expressed from the second copy of the *lac* operon will bind to both operators, blocking expression of the *lac* structural genes until inducer is added to remove the repressor. This leads us to the next question:

Can RNA polymerase express the *lac* structural genes once repression is lifted?

This is a simple question that requires an examination of the *lac* promoter (*lacP*). If there is a mutation in the *lac* promoter (*lacP*$^{-}$), most likely RNA polymerase would be unable to bind to that promoter, and expression of the genes **under the control of that mutant promoter only** would not occur. If the *lac* promoter is wild type (*lacP*$^{+}$), RNA polymerase would bind in the absence of the repressor protein and mRNA would be made **from that copy of the *lac* operon**, although it is important to realize that just because mRNA is synthesized does not mean that functional protein will be produced. Both copies of the *lac* operon in our example are under the control of wild-type promoters, so RNA polymerase will be able to bind to the promoters after the *lacI* repressor protein is removed by inducer. The final question is as follows:

Will functional β-galactosidase be produced?

If there is a mutation in the *lacZ* gene (*lacZ*$^{-}$), functional β-galactosidase will not be made, regardless of whether lactose is present or not. This is the final check point. Only one functional *lacZ* gene is needed, but it *must* be part of a *lac* operon with a functional promoter. Both copies of the *lacZ* gene in this example fit this description, so functional β-galactosidase will be produced in this cell as long as lactose is present to remove the repressor (the actual inducer that binds to the *lacI* repressor protein is allolactose, a derivative of lactose that is automatically produced in the cell when lactose is present).

Another consideration is whether functional lactose permease (the product of the *lacY* gene) is present. If lactose permease is not present (e.g., in a *lacY*$^{-}$ cell), lactose cannot be brought into the cell, so the inducer (allolactose) would not be produced and the *lacI* repressor protein would not be removed from the operator sequence. In this case, even if lactose was present in the external medium, β-galactosidase would not be expressed.

2. Under what conditions would a cell with the following merodiploid genotype express functional β-galactosidase?

$lacI^+\ lacO^+\ lacP^-\ lacZ^+\ Y^+\ A^+/lacI^+\ lacO^C\ lacP^+\ lacZ^-\ Y^+\ A^+$

Is there active *lacI* repressor protein in the cell?

Both copies of *lacI* are wild type, so *lacI* repressor protein will be present in the cell.

Can the *lacI* repressor protein bind to both operator sites?

One of the operator sequences (copy 1 of the *lac* operon) is wild type, whereas the other copy of the *lac* operator (copy 2) has a constitutive mutation ($lacO^C$). This means that the *lacI* repressor protein will only be able to bind to the operator sequence on copy 1. Expression of the *lac* genes on copy 2 will be constitutive or continuous in the absence of lactose, provided that RNA polymerase can bind to the promoter on that copy of the *lac* operon.

Can RNA polymerase express the *lac* structural genes once repression is lifted?

There is a wild-type promoter ($lacP^+$) on copy 2 only, the same copy with the constitutive operator. Therefore, RNA polymerase will be able to bind to this copy of the *lac* operon and this binding will be independent of lactose because of the constitutive operator.

Will functional β-galactosidase be produced?

We know that no functional β-galactosidase can be produced from copy 1 because of the $lacP^-$ mutation, so we have only to look at copy 2 to address this final question. Expression of the *lac* genes will be constitutive from copy 2 because of the $lacO^C$ mutation, but no **functional** β-galactosidase will be produced from this copy because there is a mutation in the *lacZ* gene. Therefore, this merodiploid will be unable to produce **functional** β-galactosidase, regardless of whether lactose is present or not.

3. Would a cell with the following merodiploid genotype be able to make functional β-galactosidase in the presence of lactose?

$lacI^S\ lacO^+\ lacP^+\ lacZ^+\ Y^+\ A^+/lacI^+\ lacO^+\ lacP^+\ lacZ^+\ Y^+\ A^+$

Is there active *lacI* repressor protein in the cell?

This cell carries a different mutation than the typical $lacI^-$ mutation. $lacI^S$ indicates a "*s*uper-repressor" mutant that produces a repressor protein that is capable of binding to the operator but is incapable of being removed by the inducer. By definition, "active" *lacI* repressor protein is capable of binding to the operator, so, given this definition, both copies of the *lac* operon shown above can produce active *lacI* repressor protein.

Can the *lacI* repressor protein bind to both operator sites in the cell?

Both operators are wild type in this merodiploid, so the wild-type *lacI* repressor protein and the super-repressor *lacI* protein will compete for binding to the operators. In the absence of lactose, repression will not be lifted, and no β-galactosidase will be produced. What happens when lactose is added? Lactose (upon conversion to allolactose) will remove all wild-type *lacI* repressor protein from operators to which it is bound, but the inducer (allolactose) will not be able to "deactivate" the super-repressor. Therefore, as operators are vacated by inactivated wild-type *lacI* repressor protein, they will be occupied by super-repressor protein until both operators in the cell are permanently blocked. Therefore, β-galactosidase will not be efficiently produced in this merodiploid in either the presence or absence of lactose.

Chapter Test

True/False

1. The fastest way for a cell to remove an enzyme activity it no longer needs is to decrease the rate of transcription of that enzyme.
2. cAMP acts as the inducer of the *lac* operon.
3. The tryptophan operon is off in the absence of tryptophan to conserve resources when the enzymes are not needed.
4. The function of the *lacA* transacetylase protein is to transport lactose into the cell.
5. Mechanisms that regulate each step from DNA to protein activity exist in bacteria.

Short Answer

6. An *E. coli* mutant is isolated that efficiently metabolizes glucose but is unable to efficiently metabolize a number of other sugars, including lactose, arabinose, and maltose. The mutation is most likely to be in one of which two genes?
7. Briefly describe how expression of the *trp* operon genes is regulated.
8. The alteration in an enzyme whose shape is changed by the binding of a small molecule to a site different from the active site is called ___________.
9. Contrast repressible and inducible operons.
10. Describe two mutant conditions under which an *E. coli* cell would exhibit constitutive expression of β-galactosidase in the absence of lactose.

Multiple Choice

11. Which one of the following activity curves most accurately depicts the induction of β-galactosidase by lactose in an *E. coli* culture containing no other carbon (sugar) source?

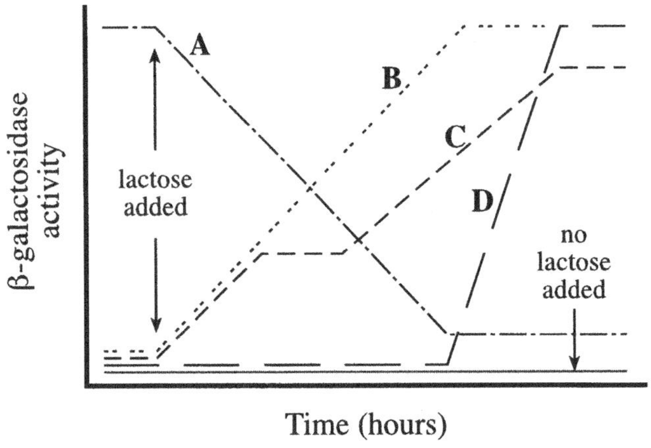

a. Plot A
b. Plot B
c. Plot C

d. Plot D
e. None of the above

12. Which one of the activity curves from the previous question would most accurately depict the induction of β-galactosidase if lactose and glucose were simultaneously added to the culture?
    a. Plot A
    b. Plot B
    c. Plot C
    d. Plot D
    e. None of the above plots

13. Which one of the following statements accurately describes the *lacI* repressor gene?
    a. The expression of the *lacI* repressor gene depends on the adjacent *lac* operator sequence.
    b. The *lacI* repressor gene is under the control of the same promoter as the *lac* operon genes.
    c. The *lacI* repressor gene produces a protein that can only bind to the *lac* operator sequence if the operator is immediately adjacent to the repressor gene.
    d. The *lacI* repressor gene gives rise to a protein which can diffuse around the cell and bind to remote *lac* operator sequences.

14. Which of the following statements accurately describes the operator for the *lac* operon?
    a. The operator produces a protein that can only bind to the *lac* promoter if the promoter is immediately adjacent to the operator.
    b. The operator gives rise to a protein that can diffuse around the cell and influence binding to remote *lac* promoters.
    c. The operator is a DNA binding site that does not give rise to a product and therefore can only influence the *lac* promoter of which it is a part.
    d. The operator is a DNA binding site and binding of the *lac* repressor to the operator can only occur in the absence of glucose.

15. Merodiploids (or partial diploids) are cells that contain a second copy of some of the genes present in their genomes. *E. coli* cells with which one of the following merodiploid *lac* genotypes would produce active β-galactosidase in the *absence* of lactose?
    a. $lacI^+\ lacO^+\ lacZ^+ / lacI^+\ lacO^+\ lacZ^+$
    b. $lacI^-\ lacO^+\ lacZ^+ / lacI^+\ lacO^+\ lacZ^-$
    c. $lacI^+\ lacO^+\ lacZ^+ / lacI^-\ lacO^+\ lacZ^-$
    d. $lacI^-\ lacO^C\ lacZ^- / lacI^+\ lacO^+\ lacZ^+$
    e. $lacI^-\ lacO^+\ lacZ^- / lacI^+\ lacO^C\ lacZ^+$

Essay

16. Bacterial genes may be expressed constitutively or may be inducible or repressible. Predict the category into which each of the following would most likely fit:
    a. A gene that codes for RNA polymerase
    b. A gene that codes for an enzyme required to break down a sugar
    c. A gene that codes for an enzyme used in the synthesis of an essential amino acid such as tryptophan

Chapter Test Answers

1. **False**
2. **False**
3. **False**
4. **False**
5. **True**
6. The gene for adenylyl cyclase (the enzyme that makes cAMP from ATP) or the gene for the CRP.
7. In the absence of tryptophan, the repressor alone cannot bind to the operator region so the *trp* genes are expressed. In the presence of tryptophan, tryptophan itself acts as a corepressor and binds to the repressor protein, activating it and allowing it to bind to the *trp* operator region, thereby turning off expression of the *trp* genes.
8. Allosteric modification
9. In a repressible system such as the *trp* operon, synthesis of tryptophan is turned off in response to the external availability of tryptophan. In an inducible system such as the *lac* operon, expression of the *lac* genes is turned on in response to the external availability of lactose.
10. If either the *lac* repressor was mutated such that it was no longer able to bind to the operator region or if the operator region itself was altered such that the *lac* repressor could not bind to it, expression of *lac* genes would be constant even in the absence of lactose.
11. **b** 12. **d** 13. **d** 14. **c**
15. **e.** Because the *lacI* repressor protein is a diffusible protein product, only one functional copy is required for negative regulation, provided that all operator sites are capable of being bound by the repressor (i.e., that they are *lacO*$^+$). In the wild-type situation, β-galactosidase is not expressed in the absence of lactose, so a is incorrect. In b and c, there is a mutation in one copy of *lacI*, but regulation is still intact because the repressor produced by the second copy of *lacI* which is wild type (*lacI*$^+$) will diffuse through the cell and block *both* operator sites. In d, no repressor can bind to the first copy of the operator site because it is mutated, so expression of genes on that copy only will be constitutive. However, there is a mutation in the *lacZ* immediately adjacent to this constitutive (mutant) operator so active β-galactosidase cannot be produced from this copy of the *lac* operon. Regulation is intact on the second copy, so no β-galactosidase will be produced in the absence of lactose. e is the correct answer. Functional repressor protein is found in this cell, but it cannot bind to the constitutive operator mutation (O^C) in the second copy of the *lac* operon, leading to constant expression of the genes on that copy. Because this second copy includes a wild-type copy of *lacZ*, β-galactosidase will be produced even in the absence of lactose.
16. **a.** Constitutive. Living cells are constantly synthesizing proteins and so would need a constant supply of RNA polymerase for transcription. For this reason, RNA polymerase would be expressed constitutively. b. Inducible. An enzyme required to break down a sugar would only be necessary when that sugar is present. Therefore, expression of a gene encoding this type of enzyme would be blocked (or repressed) until the enzyme is needed,

for example, when the sugar is present. Under these conditions, the repression is lifted and the gene is expressed. This situation is best described as inducible, and is characterized by the *lac* operon in *E. coli*. c. Repressible. Proteins cannot be synthesized without the constituent amino acids. For this reason, the cell must have a constant supply of amino acids at all times, so if a particular amino acid is not available in the surrounding medium, it has to be synthesized by the bacterium itself. On the other hand, for a bacterium to continue to synthesize an amino acid if it becomes available in the external medium would be wasteful. Expression of such a gene is said to be repressible, meaning that it is turned on by default but can be turned off if the organism is presented with the gene product.

Check Your Performance

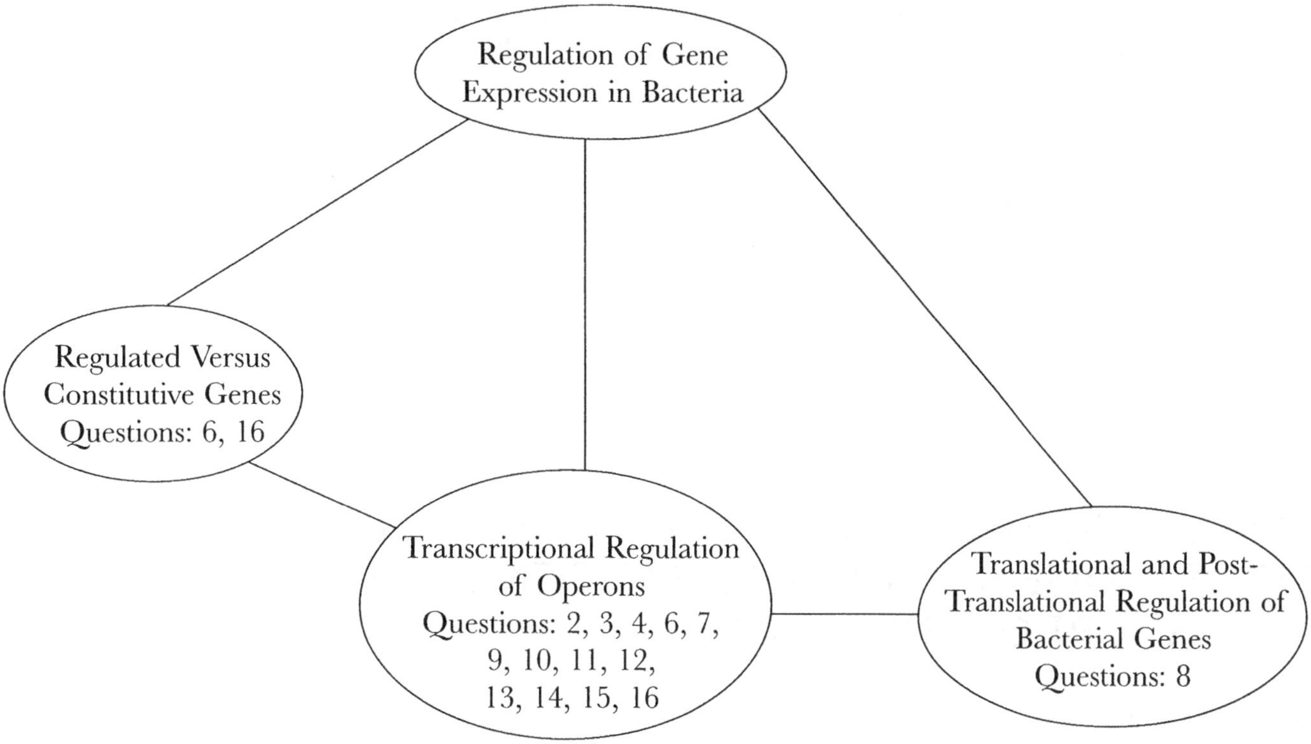

Note the number of questions in each grouping that you got wrong on the chapter test. Identify where you need further review and go back to relevant parts of this chapter.

CHAPTER 6

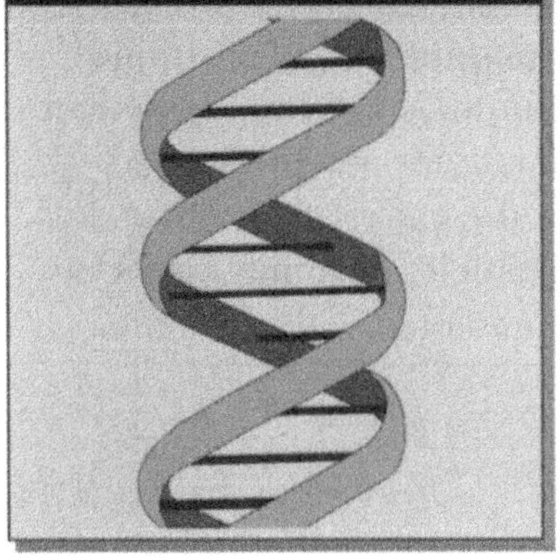

Regulation of Gene Expression in Eukaryotes

Most eukaryotes are multicellular organisms containing many different cell types (more than 200 cell types exist in the human body). However, with few exceptions, the same genetic information is present in every cell; the different cell types arise from the differential expression of an identical genome. Moreover, most eukaryotic cells are exposed to a fairly constant internal environment. Therefore, in contrast to the unicellular bacteria where the primary role of gene regulation is to allow the cell to respond to changes in its environment, the primary role of gene regulation in eukaryotes is to ensure that the right gene is expressed in the right cell at the right time during development.

ESSENTIAL BACKGROUND

- **Structure of eukaryotic chromosomes**
- **Transcription and translation of eukaryotic genes**
- **RNA processing: role of small nuclear ribonucleoproteins**

TOPIC 1: REGULATORY PROTEINS AFFECT RNA POLYMERASE BINDING AND EFFICIENCY OF TRANSCRIPTION INITIATION

KEY POINTS

✓ *Why is gene regulation in eukaryotes more complex than gene regulation in bacteria?*

✓ *How do eukaryotes ensure that the right genes are turned on in the right cells?*

✓ *Do all the gene regulatory mechanisms in eukaryotes serve to enhance gene expression?*

✓ *How do eukaryotic cells regulate the expression of genes distributed throughout the genome that need to be turned on or off at the same time?*

Gene regulation in eukaryotes is more complicated than gene regulation in bacteria for a number of reasons. First, most eukaryotic cells contain much more DNA and many more genes that need to be regulated than bacterial cells. For example, a typical human cell contains approximately 6,000 million base pairs of DNA and an estimated 100,000 genes compared with the 4.7 million base pairs of DNA and approximately 4,000 genes present in an *Escherichia coli* cell. Second, in both eukaryotes and bacteria, the DNA is wrapped around proteins to form a more compact structure called chromatin, which itself undergoes various degrees of packaging to further condense the genome. Packaging of eukaryotic chromatin is considerably more complex than bacterial chromatin and extensive compaction restricts access of the transcription apparatus to the DNA. Third, in eukaryotes, the genetic material is contained within the nucleus (and to a lesser extent in the mitochondria and chloroplasts), whereas most of the protein-synthesizing apparatus is located in the cytoplasm. Therefore, in contrast to bacteria, where the genetic material and all the components needed to express that genetic material are located in the same cellular compartment (the cytoplasm), in eukaryotes, the mRNAs that are synthesized in the nucleus need to be transported into the cytoplasm to be translated. This adds an extra step at which gene expression can be regulated.

Fourth, in contrast to bacterial genomes where most DNA codes for proteins (or for transfer RNA [tRNA] and ribosomal RNA [rRNA]), most eukaryotic genomes contain enormous amounts of DNA that does not encode protein or RNA. This noncoding DNA is found between genes but can also occur within genes (introns) where it interrupts the coding sequence and represents a further complication in gene expression. Finally, in contrast to bacterial genomes, where functionally related genes are simultaneously expressed from operons, related eukaryotic genes are often scattered throughout the genome. Most eukaryotic genomes also contain many DNA sequences that are repeated hundreds, thousands, or even millions of times. Most of these highly repetitive sequences are not transcribed but some are, for example, the numerous genes for rRNA and tRNA. Coordinating the expression of scattered functionally related or repetitive genes poses yet another complication in eukaryotic gene expression.

As with bacteria, gene regulation can and does occur at each step in the pathway from DNA to protein. The greater complexity of eukaryotic genomes offers opportunities for controlling gene expression at additional steps (e.g., at the posttranscriptional level) but, as with bacteria, whether a gene is expressed or not and how efficiently it is expressed is most often determined at the level of transcription itself.

TRANSCRIPTIONAL REGULATION

"General" Regulatory Proteins Are Essential for the Binding of Eukaryotic RNA Polymerases

In a multicellular organism with many different cell types, different genes are expressed in different cells at different times during development, and most of this control is exerted at the level of transcription.

The transcription apparatus in bacteria is relatively simple and comprises only one type of RNA polymerase that is responsible for the synthesis of all three types of RNA (messenger RNA [mRNA], tRNA, and rRNA). By comparison, the transcription apparatus in eukaryotes is surprisingly complex. Thus, as described in Chapter 2, eukaryotes contain three different "gene type"-specific RNA polymerases (where gene type refers to the type of RNA that the gene encodes), each of which recognizes a different gene type-specific promoter. In addition, the eukaryotic transcription apparatus is absolutely dependent on a large number of regulatory proteins that when bound to specific regulatory sequences, either enhance or inhibit RNA polymerase binding and therefore gene expression.

Thus, whereas *E. coli* RNA polymerase by itself can recognize the promoter and initiate transcription, by contrast, the eukaryotic RNA polymerases by themselves cannot bind to their promoters—they can bind only after various regulatory proteins called **general or basal transcription factors** have assembled there. In the case of eukaryotic RNA polymerase II, which transcribes the protein-encoding genes, the general transcription factors bind to the highly conserved TATA box that is located approximately 30 base pairs upstream of the transcription initiation site and, together with the RNA polymerase, form the minimal complex necessary for promoter recognition and transcription initiation (**Figure 6.1**).

Right Combination of General and Cell-specific Regulatory Proteins Is Necessary for High Level Expression of Cell-specific Genes

The general transcription factors are essential for the transcription of all eukaryotic genes transcribed by RNA polymerase II and are found in all the cells of an organism. Other regulatory proteins are only found in certain tissues and they recognize sequences that are specific to only a few eukaryotic genes. These cell-specific regulatory proteins (also called **cell-specific transcription factors**) play an important role in cellular differentiation—the specialization of cells during development.

Thus, in addition to the promoter sequences that bind general transcription factors, three other regions of eukaryotic DNA that bind cell-specific regulatory proteins that influence RNA polymerase binding have been identified. Some of these regulatory sequences are located just upstream of the promoter (the so-called **regulator regions** or **upstream promoter elements**), whereas others, called **enhancers**, may be thousands of base pairs away from the promoter and may even be downstream of the gene. Both the upstream promoter elements and enhancer regions bind regulatory proteins that strongly stimulate transcription initiation. How enhancers with their bound proteins can influence transcription from a distance is poorly understood. However, it is known that, in some cases, the regulatory protein bound to an enhancer sequence stimulates transcription by directly contacting the RNA polymerase and general transcription factors by looping out the intervening DNA (**Figure 6.2**).

In addition to regulatory proteins that enhance RNA polymerase binding, there are regulatory proteins that inhibit transcription. These so-called **repressor proteins** bind to sequences called **silencers** that are thought to be analogous to enhancers except that they have the reverse effect, probably disrupting interactions within the transcription apparatus assembled at the promoter.

Enhancers and upstream promoter elements may not be needed for the expression of "**housekeeping genes**," so-called because they are expressed in virtually all cells at all times, for example, genes whose products are involved in cellular respiration or protein synthesis. By contrast, enhancers and upstream promoter elements seem to be associated with most genes whose expression is "regulated," and the right combination of general and cell-specific regulatory proteins is necessary for high levels of expression of these regulated genes. Because eukaryotic gene regulatory proteins can control transcription when bound to DNA many nucleotides away from the promoter, the regulatory sequences that control the expression of a gene can be spread over long stretches of DNA and most eukaryotic genes are actually controlled by numerous regulatory sites for both positively and negatively acting regulatory proteins. Thus, most eukaryotic gene regulatory proteins work as part of a "committee," all of which are necessary to express the gene at the required level, in the right cell, and at the right time. For example, in immature red blood

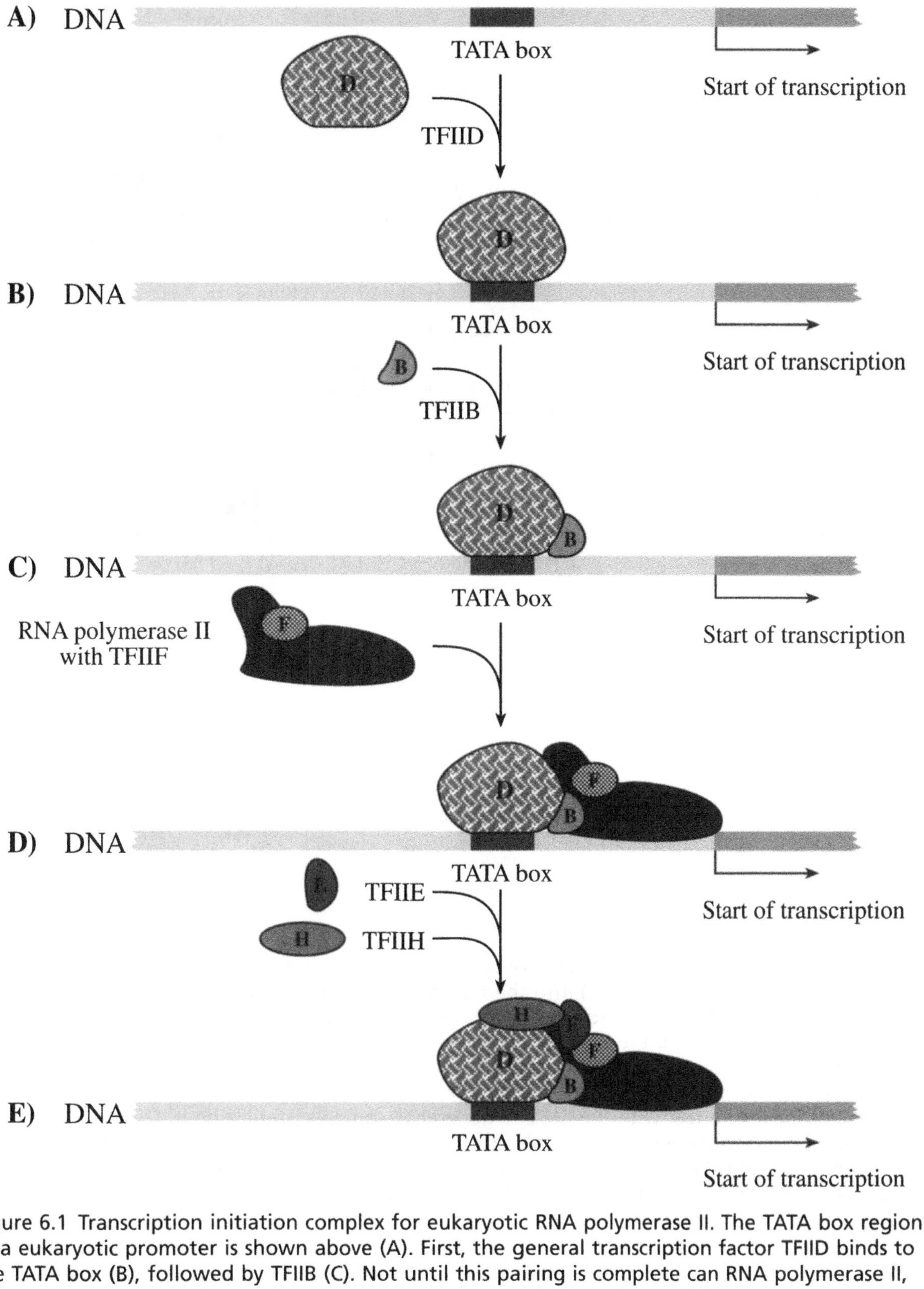

Figure 6.1 Transcription initiation complex for eukaryotic RNA polymerase II. The TATA box region of a eukaryotic promoter is shown above (A). First, the general transcription factor TFIID binds to the TATA box (B), followed by TFIIB (C). Not until this pairing is complete can RNA polymerase II, which is complexed with TFIIF, bind to the DNA (D). The binding of two additional general transcription factors, TFIIE and TFIIH (E), is necessary before RNA polymerase II can initiate transcription.

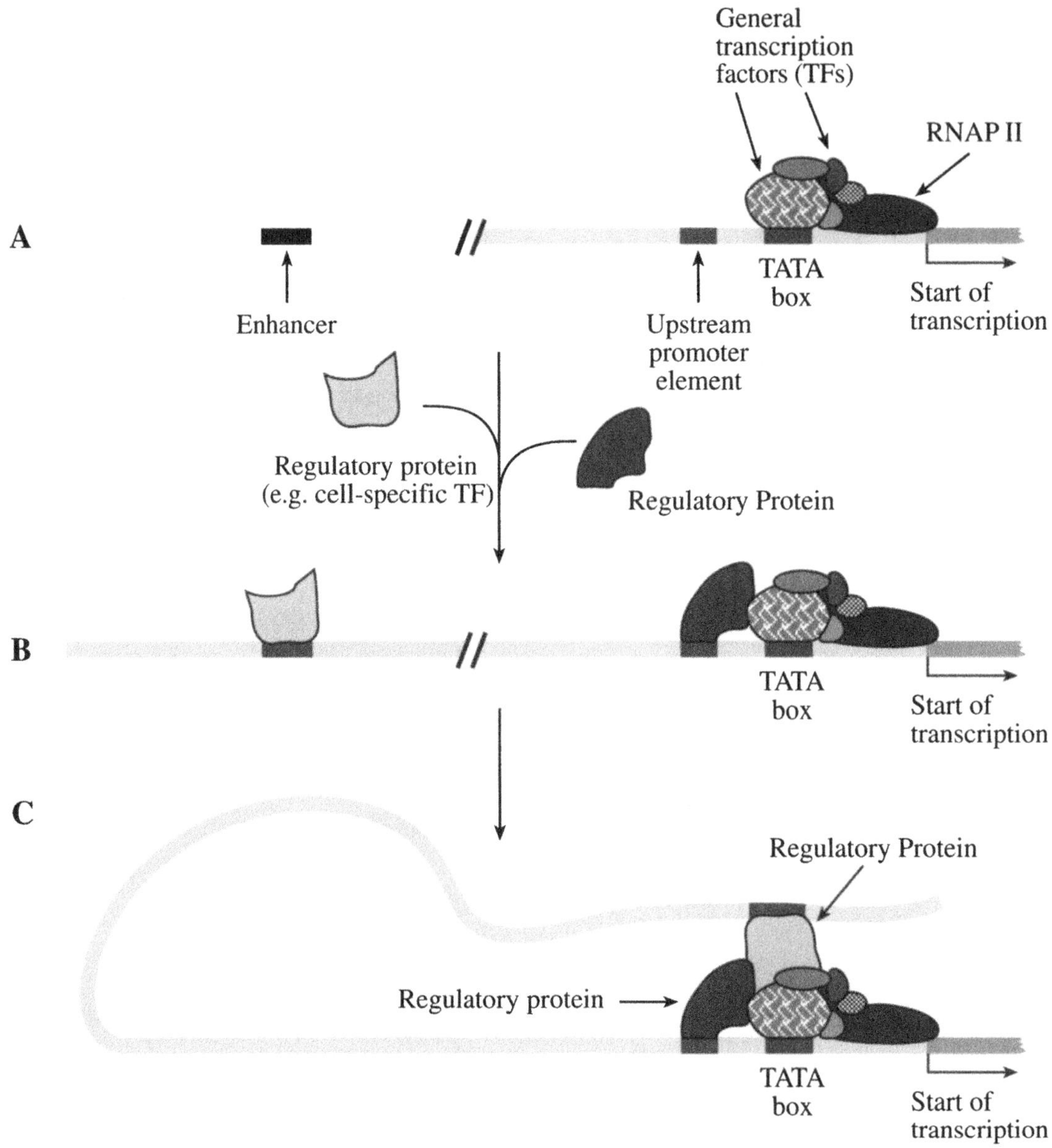

Figure 6.2 Enhancers and activator proteins. A. The assembly of general transcription factors (TFs) and RNA polymerase II (RNAP II) into the transcription initiation complex described in Figure 6.1 is shown, together with two additional regulatory regions. Upstream promoter elements are generally found close to the TATA promoter sequence, whereas enhancers can be found great distances from the start of transcription (as indicated by the "//" in the DNA). B. Gene regulatory proteins bind to these regulatory regions. The regulatory protein bound to the upstream promoter element makes contact with the RNAP II/general TF complex, stabilizing it. C. The regulatory protein bound to the enhancer can also contact the RNAP II/general TF complex by bending of the intervening DNA. The right combination of general and cell-specific transcription factors is necessary for efficient expression of cell-specific genes.

cells that make large amounts of the protein β-globin, 13 different regulatory proteins are required for efficient expression of the β-globin gene. Some of these regulatory proteins are found in many cell types including the closely related white blood cells, whereas others (e.g., GATA-1) are present only in red blood cell precursors. A unique combination of regulatory proteins is thought to be responsible for the specific expression of the β-globin gene in red blood cells.

COORDINATE REGULATION OF EUKARYOTIC GENES

As explained in Chapter 5, in bacteria, genes whose products are involved in the same pathway are often clustered together in operons where they are expressed simultaneously into a single polycistronic mRNA. In eukaryotes, on the other hand, related genes are not clustered into operons; instead, they exist as solitary genes that are often scattered throughout the genome. How do eukaryotic cells regulate the expression of distributed genes that need to be turned on or off at the same time?

Coordinate regulation of related eukaryotic genes can be achieved if these genes share the same regulatory sequences (enhancers and upstream promoter elements) and bind the same regulatory proteins. Although, as just mentioned, control of gene expression often requires a team of proteins working together, a single regulatory protein can still be decisive in switching any particular gene on or off simply by completing the combination needed to activate or repress expression of that gene. The ability to switch genes on or off using just one protein is one of the means by which eukaryotic cells differentiate during embryonic development. For example, the mature muscle cell produces a large number of characteristic proteins and the genes encoding these muscle-specific proteins are all switched on coordinately by regulatory proteins that bind to sites present in all their regulatory regions. Interestingly, when the gene for one of these regulatory proteins, MyoD, was introduced into fibroblasts, which are derived from the same class of embryonic cells as muscle cells, the fibroblasts differentiated into muscle cells. The ability of MyoD alone to convert a fibroblast into a muscle cell suggests that fibroblasts must contain all the other regulatory proteins required for the coordinated expression of the muscle-specific genes. The addition of MyoD completes the unique combination that directs the cells to become muscle as opposed to skin cells. The conversion of a fibroblast to a cell by a single gene regulatory protein is a dramatic demonstration that the differences between cell types are produced by differences in gene expression.

Topic Test 1: Regulatory Proteins Affect RNA Polymerase Binding and Efficiency of Transcription Initiation

True/False

1. A transcription factor is composed of amino acids; an enhancer is composed of nucleotides.

2. Cell-specific gene expression is brought about by the binding of cell-specific regulatory proteins to cell-specific enhancer sequences.

3. Unlike bacterial gene regulatory mechanisms that most often function at the level of transcription, most eukaryotic gene regulation occurs before transcription.

Multiple Choice

4. Your muscle cells and skin cells look different because:
   a. different genes are present in each of these cell types.
   b. the same proteins are produced in each of these cell types but they perform different functions in each kind of cell.
   c. different proteins are produced in each of these cell types.
   d. the same proteins are produced in each of these cell types but different proteins are activated in each kind of cell.
   e. All of the above.

5. All of the following are involved in the transcriptional control of gene expression (i.e., they play a role in increasing or decreasing the production of RNA by RNA polymerase), except
   a. enhancer sequences.
   b. transcription factors.
   c. small nuclear ribonucleoproteins.
   d. operator sequences.
   e. repressor proteins.

6. Which one of the following statements about the transcription of eukaryotic genes is correct? Transcription of eukaryotic genes
   a. is carried out by a single RNA polymerase.
   b. produces mRNA molecules only.
   c. occurs in the same cellular compartment as translation.
   d. is absolutely dependent upon accessory proteins.
   e. All of the above

7. Gene X is expressed only in liver cells and gene Y is expressed only in pancreatic cells. You want to express gene X in pancreatic cells. You could do this by (Hint: there may be more than one answer to this question)
   a. moving gene X into pancreatic cells.
   b. replacing the regulatory regions (e.g., enhancers and upstream promoter elements) for gene Y with those for gene X in pancreatic cells.
   c. replacing the regulatory regions for gene X with those for gene Y in liver cells.
   d. replacing the regulatory regions for gene X with those for gene Y in pancreatic cells.
   e. isolating the genes for the liver-specific regulatory proteins that bind to the regulatory regions of gene X in liver cells and expressing them in pancreatic cells.
   f. isolating the genes for the pancreas-specific regulatory proteins that bind to the regulatory regions of gene Y in pancreatic cells and expressing them in liver cells.

Short Answer

8. What strategy for gene regulation do bacteria and eukaryotes appear to have in common?

9. If eukaryotic posttranscriptional processing was suddenly stopped, which type of RNA (if any) would be affected?

10. ____________ bind at the TATA box of eukaryotic promoters and enhance the binding of RNA polymerase II; negatively acting repressor proteins bind to ____________ and

decrease gene expression levels by disrupting interactions within the transcription initiation apparatus.

Topic Test 1: Answers

1. **True.** Transcription factors are proteins that bind nucleotide sequences and regulate gene expression, and proteins consist of amino acids. Enhancers are DNA sequences that are bound by regulatory proteins such as cell-specific transcription factors, and because they are DNA sequences, they are composed of nucleotides.
2. **False.** Enhancer sequences are part of the DNA sequence, and, with few exceptions, the same genetic material (DNA) is found in each and every cell. It is one or a few of the regulatory proteins that bind to the enhancer sequences that are usually cell specific.
3. **False.** It is correct that there are more pretranscriptional regulatory mechanisms in eukaryotes than in bacteria because of the greater complexity of the eukaryotic genome, but most regulation in eukaryotes still occurs at the level of transcription.
4. **c.** Each somatic cell in your body contains the same genetic information, both in number and identity of genes. This means that a is not correct. b and d are incorrect because each specialized cell type produces different cell-specific proteins to accomplish the tasks required by that cell type. This leaves c as the only correct answer.
5. **c.** Enhancer sequences (in conjunction with cell-specific regulatory proteins) and general transcription factors enhance the binding of RNA polymerase to a eukaryotic promoter sequence, thereby increasing the frequency of transcription. Likewise, operator sequences and the repressor proteins that bind to them inhibit the binding of bacterial RNA polymerase, thereby also affecting transcription. Only small nuclear ribonucleoproteins, which are responsible for the removal of introns, operate at the posttranscriptional level.
6. **d.** There are three eukaryotic RNA polymerases, as opposed to one in bacteria. This makes a incorrect. b is not correct because transcription is the synthesis of RNA from a DNA template, and there are three types of RNA (mRNA, tRNA, and rRNA). Finally, c is incorrect because transcription in eukaryotes takes place in the nucleus, whereas translation occurs in the cytoplasm. Only d is correct, because eukaryotic RNA polymerases cannot bind to their promoter sequences in the absence of general transcription factors.
7. **d and e.** Moving gene X into pancreatic cells will not help because gene X is already present in pancreatic cells, it just is not expressed. Replacing the regulatory regions for gene Y with those for gene X in pancreatic cells would serve to turn off gene Y in pancreatic cells because the pancreas-specific regulatory proteins that are responsible for turning on gene Y in pancreatic cells cannot recognize the regulatory regions of gene X (if they did, then gene X would be expressed in the pancreas). Replacing the regulatory regions for gene X with those for gene Y would allow gene X to be turned on by the same pancreas-specific regulatory proteins that turn on gene Y but if this is done in liver cells, as the statement suggests, then the pancreas-specific regulatory proteins would not be present, so no expression of gene X would occur. Isolating the genes for the pancreas-specific regulatory proteins that bind to the regulatory regions of gene Y in pancreatic

cells and expressing them in liver cells would succeed in turning on gene Y in liver cells and in pancreatic cells, which is not what the question asks for. This eliminates a, b, c, and f, respectively. Replacing the regulatory regions for gene X with those for gene Y in pancreatic cells would provide gene X with the necessary sequences to be activated by the same pancreas-specific regulatory proteins that activate gene Y. In other words, bringing regulatory regions and the specific regulatory proteins that recognize them together into the same cell will result in expression of that gene. Similarly, isolating the genes for the liver-specific regulatory proteins that bind to the regulatory regions of gene X in liver cells and expressing them in pancreatic cells would provide the copy of gene X present in pancreatic cells with its necessary regulatory proteins so expression would occur. This explains why options d and e are both correct.

8. Both bacteria and eukaryotes use regulatory proteins that bind to specific DNA sequences and affect the levels of gene expression. In eukaryotes, these regulatory proteins can take the form of general transcription factors that bind to specific promoter sequences or cell-specific regulatory proteins that bind to upstream promoter elements, enhancers, or silencers; in bacteria, repressor proteins such as the *lacI* repressor protein, which binds to the operator, or global regulators like the cyclic AMP–CRP complex, which binds to specific promoter sequences, fill a similar role.
9. All three types of RNA (mRNA, tRNA, and rRNA) could be affected. Introns are found in all three gene types (i.e., genes encoding mRNAs, tRNAs, and rRNAs), and these introns are removed from the primary transcript by posttranscriptional RNA processing. Moreover, all three RNAs are further processed: mRNA by the addition of 5′ cap and 3′ tail structures and rRNAs and tRNAs by the removal of additional nucleotides. All these processing steps occur at the posttranscriptional level.
10. **General transcription factors** bind at the TATA box of eukaryotic promoters and enhance the binding of RNA polymerase II; negatively acting repressor proteins bind to **silencers** and decrease gene expression levels by disrupting interactions within the transcription initiation apparatus.

TOPIC 2: OTHER MECHANISMS FOR THE REGULATION OF EUKARYOTIC GENES

Key Points

✓ *Why do most eukaryotic genomes contain methylated bases?*

✓ *How can different proteins be produced from a single eukaryotic gene?*

✓ *What signals promote the binding of specific gene regulatory proteins to their regulatory sequences?*

CHROMATIN MODIFICATIONS

Initiation of transcription in eukaryotic cells is also influenced by the packing of DNA into chromatin. For example, it is known that nucleosomes can inhibit the initiation of transcription if they are positioned over a promoter, most likely by preventing the general transcription factors and RNA polymerase from assembling on the DNA.

Key enzymes modify chromatin structure and, in so doing, make genes more or less accessible to RNA polymerase. For example, **histone acetylation**, or the attachment of acetyl groups to certain amino acids of histone proteins, is thought to play a role in turning genes on. The major factors that control the condensing and decondensing of chromatin are thought to operate through the histone proteins around which the DNA is wrapped. Acetylation neutralizes the positive charge carried by the histone proteins, as a result of which they grip the negatively charged DNA less tightly. Thus, it is easier for RNA polymerase to access genes in acetylated regions of the DNA. Coincidentally, enzymes that acetylate or deacetylate histones are closely associated with the transcription factors that bind to promoter regions.

DNA methylation, on the other hand, may prevent genes from being expressed. Thus, it is has been observed that highly methylated DNA is usually not transcribed. For example, the inactivated X chromosome in mammals is highly methylated compared with actively transcribed DNA, and in many species, methylation appears to be responsible for the long-term inactivation of genes that occurs during cellular differentiation in the embryo. In particular, in mammalian cells, the methylation of C-5 in the cytosine base of 5′-CG-3′ sequences is thought to be associated with transcriptionally inactive regions of chromatin. It is now known that vertebrate cells possess a protein that binds to clusters of 5-methylcytosine and prevents transcriptional regulatory proteins from gaining access to the DNA. Methylation is viewed as having a less direct role in gene regulation than histone acetylation, being responsible not for turning genes off *per se* but for keeping genes that have already been turned off from being turned back on.

In some cases, gene expression is regulated by movement of a gene to a different location on the chromosome. For example, the yeast *Saccharomyces cerevisiae* exists in one of two mating types, a and ∝, and which mating type is expressed is controlled by the movement of the a and ∝ alleles on the chromosome. Thus, the a and ∝ alleles are usually transcriptionally silent due to the binding of a repressor protein. However, when a copy of the a or ∝ allele inserts at the MAT locus, which is the site of expression, the gene for proteins of the appropriate mating type is expressed. A change in mating type is brought about by removal of the current occupant of the MAT locus (e.g., the a allele) and insertion of a duplicate copy of the other allele (i.e., ∝). Such **DNA rearrangements** also play an important role in generating the wide variety of antibody proteins in the immune system.

Finally, increased production of some gene products is brought about by making more copies of the appropriate gene, a process referred to as **gene amplification**. As mentioned at the beginning of this chapter, some eukaryotic genes exist in many copies (e.g., the rRNA genes) but in some instances, even moderate repetition of a gene is not sufficient to satisfy the demands of the cell for that gene product. For example, the mature eggs of frogs have up to one trillion ribosomes to accommodate the massive protein synthesis that follows fertilization. How can a single cell produce enough rRNA for so many ribosomes? In the egg cell this problem is solved by selective amplification of the rRNA gene clusters (the genes for rRNAs are usually linked together in a unit) until there are more than a million copies of these clusters, compared with less than one thousand copies in the unamplified DNA. The mechanism by which genes are selectively amplified is not clearly understood but is very important because it greatly increases the amount of template available for transcription.

POSTTRANSCRIPTIONAL REGULATION

As explained in Chapter 2, the RNAs synthesized by eukaryotic RNA polymerases (so-called pre-RNAs) have to undergo various degrees of RNA processing before they can be used by the cell.

Transcriptional control processes affect the production of these pre-RNAs by RNA polymerase, most commonly by controlling access of RNA polymerase to the promoter. Posttranscriptional control processes, on the other hand, affect the production of functional RNAs from the pre-RNAs by blocking one or more of the RNA processing steps. Most eukaryotic genes are interrupted by noncoding sequences called introns. These introns are transcribed by RNA polymerase, but they have to be removed from the primary RNA transcript to produce the "mature" RNA. The splicing machinery recognizes the boundaries between introns and exons, but which RNA segments are treated as introns and which as exons can occasionally vary from one cell type to another (**Figure 6.3A**). For example, alternative intron-exon cassette selections occur in the gene sequence for the skeletal muscle protein troponin T. In the rat this gene has 18 exons, 5 of which can be omitted or used in any combination and 2 of which are mutually exclusive, resulting in 64 possible different mRNAs that can be made from a single gene. The

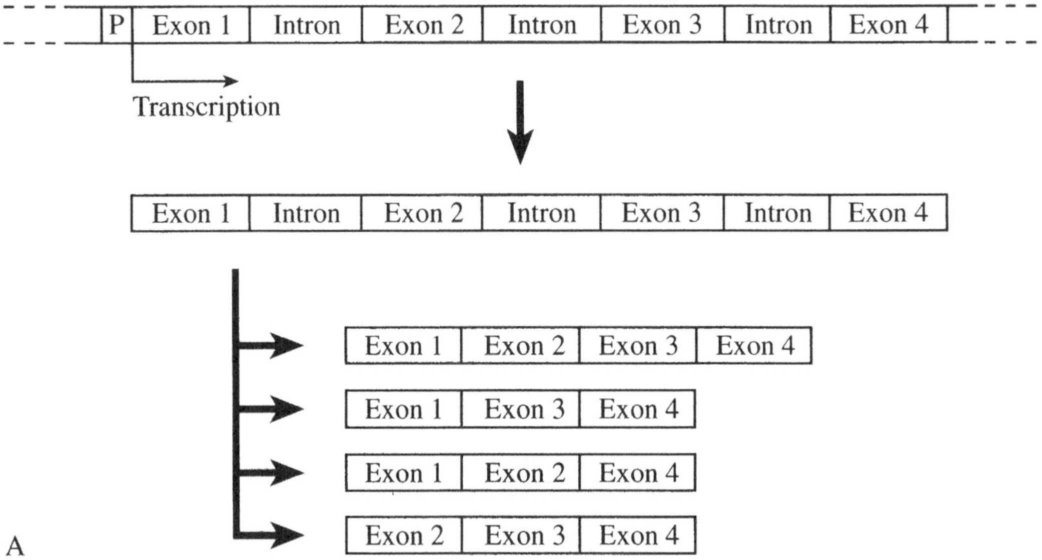

Figure 6.3 A. Principle of alternative exon cassette selection. In this example, a single primary transcript is synthesized from a single promoter. Each intron/exon junction is functional—alternative splicing is accomplished by cell-specific regulatory proteins that recognize different splice sites, resulting in different combinations of exons. The figure above shows four different exon "cassettes" that could theoretically be made from the primary transcript. All of these exon cassettes originate from the same RNA template but each would give rise to a different protein. B. Principle of alternative promoter selection. Two different primary transcripts are made from the DNA template, based on which promoter is used. If P1 is used, the primary transcript includes all introns and exons. It also includes P2, which lies between Intron 1 and Exon 2. P2 is not recognized as a functional intron/exon junction, resulting in the removal of Exon 2 in addition to the introns. This leaves a mature mRNA consisting of Exons 1, 3, and 4. If P2 is used as the promoter, the primary transcript includes only Exons 2, 3, and 4, and the introns between them. Because all intron/exon junctions are functional, splicing results in a mature mRNA consisting of Exons 2, 3, and 4. C. Principle of alternative poly(A) tail site selection. Two different primary transcripts are made from the DNA template. In one, the transcript is shortened by the addition of the poly(A) tail after Exon 3. Normal intron/exon boundaries exist throughout this transcript, resulting in a mature mRNA consisting of Exons 1, 2, and 3. In the second primary transcript, the poly(A) tail is added after Exon 4. The upstream poly(A) tail site that separates Exon 3 from the intron in front of Exon 4 is not recognized as a valid intron/exon boundary by the spliceosome, resulting in the removal of Exon 3 in addition to all of the introns. This leads to a mature mRNA consisting of Exons 1, 2, and 4. This mRNA will give rise to a different protein product than the mRNA that contains Exons 1, 2, and 3.

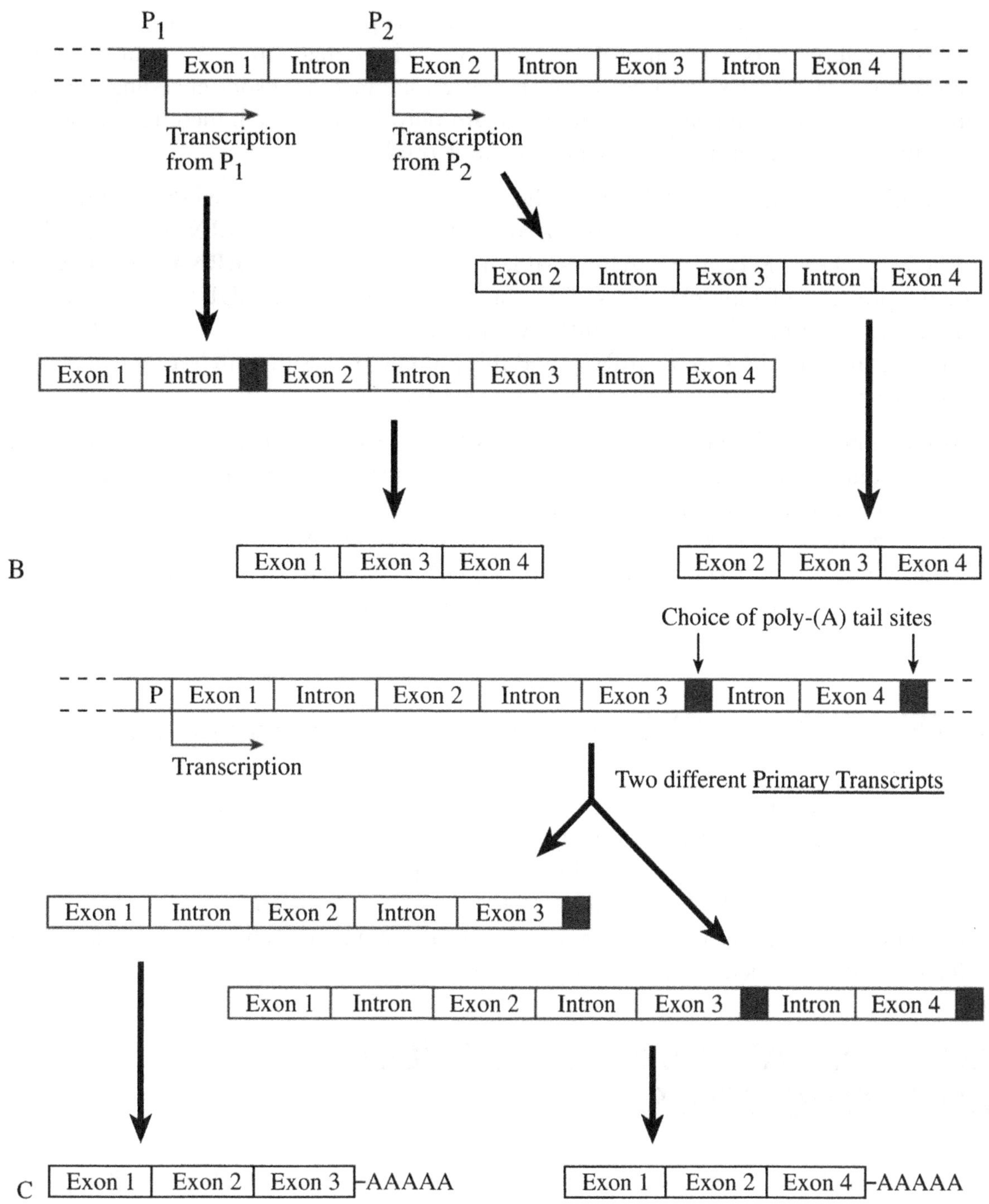

Figure 6.3 *Continued*

same mechanism is responsible for the wide variety of antibodies in the immune system. Cell-specific regulatory proteins control intron-exon choices by binding to regulatory sequences within the primary transcript. Thus, eukaryotic cells have evolved alternative modes of RNA splicing (referred to as **alternative splicing**) as a strategy to produce different proteins from a single gene. In addition, other mechanisms that lead to different proteins being produced from a single gene exist. For example, some genes have two alternative promoter sites, and yet others have alternative sites for adding the poly(A) tail, the choice between these alternative promoters or poly (A) tail sites being dependent on cell-specific regulatory proteins (Figures 6.3B and C). For example, there are two different types of the immune protein IgM, a secreted form and a membrane-bound form. Both of these forms of IgM are made from the same gene by the use of alternative tail sites. Use of a downstream poly(A) site includes exons encoding membrane-anchoring regions, whereas, when the upstream poly(A) site is used, these regions are not present and the secreted form of the immunoglobulin is produced.

Obviously, the longer an mRNA lasts, the more protein that can be synthesized from it. Therefore, mechanisms that slow down or speed up the degradation of mRNAs regulate the levels of protein synthesis. Recent research has shown that, in some cases, **mRNA stability** is influenced by the presence of multiple copies of the nucleotide sequence 5′-AUUUA-3′ in the untranslated trailer region at the 3′ end of the mRNA. In other cases, when a particular protein is in excess, the protein itself binds to the mRNA and this binding makes the mRNA more susceptible to breakdown.

As mentioned in Chapter 2, eukaryotic mRNAs are "capped" at their 5′ ends by the addition of a modified guanosine residue. Mechanisms that interfere with the 5′ cap structure of eukaryotic mRNAs have been discovered. For example, the egg cells of many organisms synthesize and store large numbers of mRNAs, but these mRNAs are not translated until after fertilization. It is now known that the stored mRNAs contain 5′ guanosine cap residues, but the caps are not properly modified until after fertilization has occurred, at which point the mRNAs can be translated.

TRANSLATIONAL REGULATION OF EUKARYOTIC GENES

Most translational control mechanisms act by blocking translation initiation. For example, in humans, production of ferritin, which is responsible for binding free iron ions, is normally shut off by the binding of a translational repressor protein called aconitase. Aconitase binds to the 5′ end of the ferritin mRNA, forming a stable loop to which ribosomes cannot bind. When the concentration of iron increases, the iron binds to aconitase causing it to dissociate from the ferritin mRNA, which can then be translated.

POSTTRANSLATIONAL REGULATION OF EUKARYOTIC GENES

Translational control processes affect the levels of protein synthesis, but so-called posttranslational control mechanisms that regulate the proteins themselves also exist. Many newly translated polypeptides do not immediately generate functional proteins as, aside from correct folding, there are usually a number of other alterations that are required for activity. For example, some proteins have to be cleaved, whereas others require chemical modifications such as the addition of sugar or phosphate groups. Regulation can occur at any one of the steps involved in modifying a protein.

Finally, different proteins have very different half-lives, and in eukaryotes, it has been discovered that the N-terminal residue plays a critical role in inherent stability. A protein that is modified or has an inherently "destabilizing" N-terminal residue becomes ubiquitinylated by the covalent linkage of molecules of the small protein ubiquitin. The ubiquitinylated protein is then digested by proteases.

ROLE OF HORMONES IN CONTROLLING GENE EXPRESSION LEVELS

Steroid and nonsteroid hormones bind to specific regulatory proteins and stimulate gene expression. In one model, a lipid-soluble steroid hormone diffuses across the plasma membrane and binds to a specific receptor protein. This results in the release of an inhibitory molecule that was associated with the receptor protein and the now-activated receptor binds to an enhancer or upstream promoter element and stimulates transcription. For example, the glucocorticoid receptor protein is a gene regulatory protein produced in liver cells that stimulates the expression of many different genes whose products are involved in the production of glucose from amino acids and other small molecules during times of starvation. For the glucocorticoid receptor protein to bind to regulatory sites in the DNA, it must first complex with a molecule of glucocorticoid steroid hormone. Some nonsteroid hormones, on the other hand, associate with receptor proteins outside of cells and, in so doing, activate intracellular pathways that in turn activate specific regulatory proteins.

Topic Test 2: Other Mechanisms for the Regulation of Eukaryotic Genes

True/False

1. Methylation plays an important role in transcriptional regulation in eukaryotes by altering the conformation of the histone proteins around which the DNA is coiled, thereby relaxing the DNA and making it easier for RNA polymerase to access promoter sequences.
2. Eukaryotes can produce different proteins from a single gene.
3. Hemoglobin consists of four globin chains and a nonprotein pigment called heme. The concentration of heme regulates the *amount* of globin protein made but has *no* effect on the synthesis or stability of mature processed globin mRNA. This is an example of translational control.

Multiple Choice

4. Which of the following is an accurate depiction of alternative splicing?

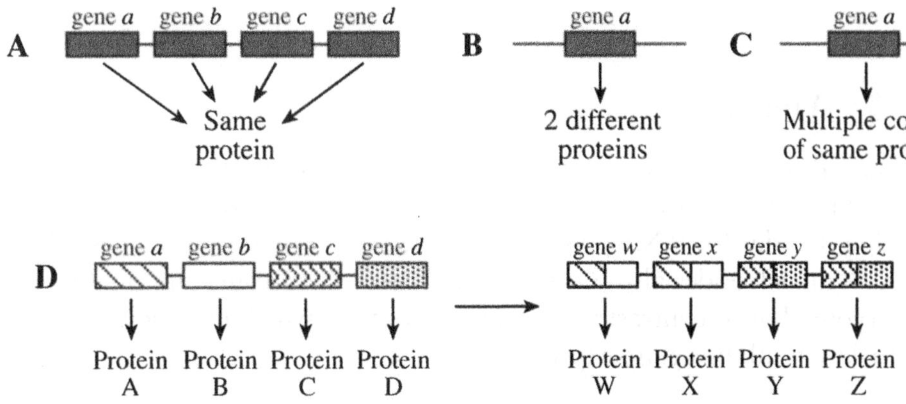

a. Figure A
b. Figure B
c. Figure C
d. Figure D
e. None of the above

5. The regulation of gene expression by mechanisms that bring about the rapid degradation of specific mRNAs is an example of
   a. transcriptional regulation.
   b. posttranscriptional regulation.
   c. translational regulation.
   d. posttranslational regulation.
   e. feedback inhibition.

6. In muscle cells, gene X contains many methylated cytosines. What might this suggest about the expression of gene X in muscle cells?
   a. Gene X is expressed only in muscle cells.
   b. Gene X is not expressed in muscle cells.
   c. Gene X is regulated at the transcriptional level.
   d. Gene X is regulated at the translational level.
   e. The fact that gene X contains many methylated cytosines in muscle cells suggests nothing about its expression in these cells.

Short Answer

7. Gene A can be alternatively spliced to give rise to two different forms of protein A, a secreted form and a membrane-bound form. The membrane-bound form of protein A is made by embryonic cells at an early stage in their differentiation, whereas the secreted form of protein A is made at a later stage of development. What is most likely responsible for ensuring that the correct form of protein A is produced at the correct time during development?

8. Enzymes called kinases phosphorylate (add phosphate groups to) certain cellular proteins. Briefly explain why kinases play an important role in the regulation of gene expression.

Topic Test 2: Answers

1. **False.** Histone acetylation (not methylation) neutralizes the positive charge carried by the histone proteins, as a result of which they grip the negatively charged DNA less tightly, making it easier for RNA polymerase to access promoter sequences. Methylation of —CG— dinucleotides also plays an important role in transcriptional regulation in eukaryotes, but in contrast to histone acetylation, which enhances gene expression, methylation inhibits gene expression.

2. **True.** This is the premise behind alternative splicing. For example, depending on which sequences in a primary transcript are treated as exons and which as introns, differing forms of a protein can be made from a single gene sequence. The use of alternative promoter sites or poly(A) tail sites also allows different proteins to be produced from essentially the same gene sequence.

3. **True.** The question specifically states that heme regulates the amount of globin but does not affect the synthesis or stability of mature processed globin mRNA. Any regulation at the transcriptional or posttranscriptional level would affect the amount of mature mRNA available for translation, which is not the case in this example. This leaves translational control or posttranslational control. Translational control would affect the *amount* of protein made from the mRNA, whereas posttranslational control would affect the *activity* of the protein made from the mRNA but not the amount of protein made. Therefore, the above statement is true.

4. **b.** Alternative splicing is a strategy for synthesizing more than one protein from a single gene, a situation illustrated in figure B. a and d are incorrect because they depict a situation where multiple genes are involved. Because an mRNA can be translated by multiple ribosomes before it is degraded, c represents the typical situation for most genes.

5. **b.** Because the mRNA is already present this form of regulation does not affect transcription. Most translational control mechanisms act by decreasing or increasing translation initiation by preventing or enhancing ribosome attachment, but in this case, the mRNAs are degraded before they can be translated. Therefore, this form of regulation occurs at the posttranscriptional level.

6. **b.** The presence of methlyated cytosines in gene X suggests that gene X has been inactivated in that cell type, because highly methylated DNA is generally transcriptionally inactive. Therefore, gene X is likely not expressed in muscle cells.

7. The production of the different forms of protein A is most likely determined by the binding of different regulatory proteins or variants of the same regulatory protein to regulatory sequences within the primary transcript for protein A. The fact that different forms of protein A are produced at different stages of development would suggest that different regulatory proteins are produced at different times during the life of the cell. For example, one possibility is that the first regulatory protein produced is itself responsible for the production of the second regulatory protein that in turn could control the production of other regulatory proteins. The sequential production of different regulatory proteins would give rise to different patterns of gene expression at different stages of development.

8. Some newly translated polypeptides do not immediately fold into functional proteins because some proteins have to be posttranslationally modified to make them active. For example, some newly translated polypeptides need to be cleaved, whereas others require chemical modifications such as the addition of sugar or phosphate groups. Therefore, kinases that posttranslationally phosphorylate proteins play an important role in gene expression and any mechanism that inhibits the kinase will prevent the protein from becoming active.

IN THE CLINIC

As we learn more about the regulation of gene activity in eukaryotes, the hope is that eventually it will be possible to produce, in the laboratory, replacement organs for the human body. Pieces of bone and skin have already been successfully cultured, and attempts are now being made to discover the transcription factors and gene regulatory proteins that control the expression of cell-specific genes that give tissues their unique identities. Once the cell-specific regulatory proteins and their regulatory sequences for a particular tissue have been characterized, the hope is that it will be possible to grow this tissue in culture. One potential problem with this technology is figuring out how the mechanical forces that are responsible for holding a tissue or organ together will be compensated for in an artificial environment. Initial studies have used rubber molds to impart the appropriate shape to the artificial tissue.

Artificial tissues will play an important role in medicine both as replacement parts (transplant tissue) and as experimental systems. For example, one of the major difficulties in gene therapy is in delivering the therapeutic drug to its target cells. Using artificial tissues, the ability of new drugs to reach the target cells could be evaluated very quickly.

DEMONSTRATION PROBLEM

You have identified a previously uncharacterized human protein that you have called Xepstan. Xepstan is the product of the *xep* gene and it is expressed only in thyroid tissue. A colleague of yours has also identified a novel protein, which is expressed only in the brain. This "brain" protein, the product of the *yor* gene, has been named Yorcal. You wish to produce Yorcal in thyroid tissue. How would you go about trying to express the *yor* gene in thyroid cells if you were told that the reason Yorcal is not expressed in these cells is because RNA polymerase is unable to bind efficiently to the promoter for the *yor* gene in thyroid tissue?

First, remember that in multicellular eukaryotes, the absence of a protein in a particular cell type does *not* indicate the absence of the gene that encodes the protein from that cell type. For example, in the case of the *xep* and *yor* genes, thyroid tissue cells contain the DNA for both *xep* and *yor*, even though only *xep* is expressed. Likewise, brain tissue cells carry the genetic material for both *xep* and *yor*, although *yor* is the only one of the two expressed in this cell type.

As described in the chapter, many eukaryotic genes are expressed on a cell-specific basis. Efficient production of a given protein is dependent on regulatory regions in the DNA that directly affect the expression of the gene for that protein. These regulatory regions are bound by proteins that enhance or inhibit the ability of RNA polymerase to transcribe the gene. Some of these proteins, called **general transcription factors**, are required for the expression of all protein-coding genes. These general transcription factors are produced in all cells and alone are sufficient to bring about the transcription of some genes that are expressed in virtually all cell types. Other regulatory proteins, however, are only found in specific cell types. These proteins, often called **cell-specific transcription factors** or **gene regulatory proteins**, are required (in addition to the general transcription factors) for the expression of cell-specific genes. For example, although both the *xep* and *yor* genes are found in thyroid tissue cells, the gene regulatory proteins that enhance expression of *yor* in brain tissue are most probably not present in thyroid cells, explaining why Yorcal is not found in the thyroid. This example illustrates the two main players involved in cell-specific gene expression: the regulatory regions within the DNA and the gene regulatory proteins that bind to those regions. If both are present, the gene is expressed. If even one is missing, the gene is not expressed.

So how can you produce a protein in thyroid tissue that is otherwise found only in brain tissue? There are two different approaches. The first is to isolate the gene regulatory proteins from brain tissue cells and introduce them into thyroid cells. The regulatory regions for *yor* already exist in thyroid tissue; it is the gene regulatory proteins that are required for the expression of *yor* that are missing. If the brain-specific gene regulatory proteins are introduced into thyroid cells, they will bind to the regulatory regions and enhance expression of *yor* in the thyroid.

It may not be possible to isolate the gene regulatory proteins from brain tissue, however. In this case, you can take advantage of the regulatory regions and gene regulatory proteins that do exist in the thyroid tissue. It is known that in the presence of thyroid-specific gene regulatory proteins, genes with the *xep* regulatory region are expressed efficiently. Therefore, if the regulatory regions for the *yor* gene present in thyroid tissue DNA were replaced with the regulatory regions for *xep*, the thyroid cells would be "tricked" into expressing *yor* in thyroid tissue. Either of these approaches would be sufficient to match up the key ingredients for cell-specific expression (regulatory regions and gene regulatory proteins), even in the "wrong" cell type, and should result in the expression of the *yor* gene in thyroid tissue cells.

Chapter Test

True/False

1. Although translational control exerts its effects faster than transcriptional control, transcriptional control is more energy efficient because it acts earlier in the path from DNA to protein.
2. Genetic regulation in both bacteria and eukaryotes occurs at multiple levels but exists primarily to adapt a cell to changes in its environment.
3. The gene for RNA polymerase II is likely to be under the control of a silencer that, together with its bound repressor protein, would act to reduce production of RNA polymerase II in cells that are less transcriptionally active than actively dividing or differentiating cells.
4. The more complex packaging of eukaryotic genomic DNA compared with bacterial genomic DNA, the compartmentalization of the eukaryotic genetic material and the presence of introns in eukaryotic genes all provide additional opportunities for genetic regulation that do not exist in most bacteria.

Multiple Choice

5. Eukaryotic RNA polymerase II by itself cannot bind to a promoter—it can bind only after the binding of regulatory proteins called
   a. general transcription factors.
   b. enhancers.
   c. cell-specific transcription factors.

d. sigma factors.
e. All of the above

6. The control of gene expression in eukaryotes is more complex than in bacteria because:
   a. in eukaryotes, different cells are specialized for different functions.
   b. eukaryotic genomes are much larger than bacterial genomes.
   c. in eukaryotes, functionally related genes are not clustered as in bacteria but instead are usually scattered throughout the genome.
   d. most eukaryotic genes are interrupted by segments of noncoding DNA.
   e. All of the above

7. Which of the following statements about transcription factors is correct?
   a. Transcription factors only bind to sequences within promoter regions of eukaryotic genes.
   b. Transcription factors are absolutely required for the transcription of eukaryotic operons.
   c. Transcription factors are absolutely required for both bacterial and eukaryotic RNA polymerases to bind to promoters.
   d. Cell-specific transcription factors play an important role in cellular differentiation in eukaryotes.
   e. All of the above

8. Which one of the following eukaryotic or bacterial DNA sequences usually gives rise to a protein product?
   a. Enhancers
   b. Promoters
   c. Operators
   d. Introns
   e. None of the above

9. Which one of the following statements about transcription in eukaryotes is incorrect?
   a. Transcription of eukaryotic genes involves three different RNA polymerases.
   b. Transcription in eukaryotes involves different genes in different cell types.
   c. Transcription of eukaryotic genes requires the binding of general transcription factors to the promoter.
   d. Most of the RNA transcripts directly produced by eukaryotic RNA polymerases are much shorter than the DNA gene sequences from which they are transcribed.
   e. Transcription of eukaryotic genes does not occur in the same cellular compartment as translation.

10. Which one of the following features of gene regulation is found in both eukaryotes and bacteria?
    a. Regulatory proteins that enhance the binding of RNA polymerase.
    b. DNA sequences that bind regulatory proteins and affect RNA polymerase binding at distant promoters.
    c. Clusters of genes under the control of a single promoter.
    d. Cell-specific proteins that regulate the transcription of cell-specific genes.
    e. All of the above

Short Answer

11. Compare and contrast enhancers and silencers.
12. How is coordinate regulation achieved in eukaryotes?
13. Most of the X chromosome's gene products are produced in equivalent amounts in both males and females. For this to occur, one of the X chromosomes is typically "shut down" in the female, a phenomenon known as X inactivation. Explain how X inactivation is thought to occur.

Essay

14. The following figure illustrates the steps involved in gene expression in eukaryotes from genes packaged in chromatin to the final functional protein:

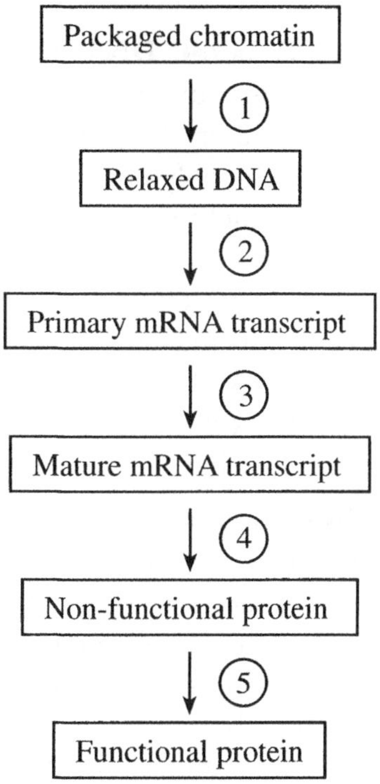

At each step labeled 1 through 5 above, describe a regulatory mechanism that affects gene expression at that step.

Chapter Test Answers

1. **True**
2. **False**
3. **False**
4. **True**
5. **a** 6. **e** 7. **d** 8. **e** 9. **d** 10. **a**
11. Enhancers and silencers are DNA sequences that can be located far from the start of transcription and both bind regulatory proteins and affect expression levels of many

eukaryotic genes. However, whereas enhancers increase expression levels of specific genes by stabilizing the complex of RNA polymerase and the general transcription factors, silencers decrease expression levels of specific genes, most likely by disrupting interactions within the transcription apparatus assembled at the promoter.

12. In eukaryotes, related genes are not clustered into operons; instead, they exist as solitary genes that are often scattered throughout the genome. However, functionally related eukaryotic genes possess similar regulatory sequences and bind the same regulatory proteins which results in them being turned on and off at the same time. Therefore, although eukaryotes do not contain operons, functionally related eukaryotic genes are also expressed in a coordinated fashion.

13. X inactivation is brought about by the methylation of bases on one of the two X chromosomes. Such methylation sterically hinders RNA polymerase from accessing the genes on the chromosome, rendering it transcriptionally silent in the heavily methylated regions.

14. **Step 1: Chromatin to Relaxed DNA.** Eukaryotic genes are less accessible than their bacterial counterparts because eukaryotic DNA is more tightly packaged than bacterial DNA. There are a number of mechanisms that affect the ability of the general transcription factors and RNA polymerase to access eukaryotic genes. For example, histone acetylation, or the attachment of acetyl groups to certain amino acids of histone proteins, is thought to play a role in turning genes on. Acetylation neutralizes the positive charge carried by the histone proteins, as a result of which they grip the negatively charged DNA less tightly. This makes it easier for RNA polymerase to access genes in acetylated regions of the DNA. The addition of methyl groups to DNA bases (particularly cytosines in 5′-CG-3′ sequences) is also thought to play an important role in gene regulation: keeping genes that have already been turned off from being turned back on.

    Step 2: Transcription. Any mechanism that affects the ability of RNA polymerase to bind to a promoter and begin synthesis of RNA works at the level of transcription. Most important at this level is the action of general transcription factors, proteins that assist the binding of RNA polymerase to the promoter and in whose absence eukaryotic RNA polymerases cannot bind and initiate transcription. All eukaryotic gene expression requires general transcription factors, but additional positive regulation is accomplished, typically in a cell-specific manner, by the binding of other regulatory proteins to specific regulatory sequences. Some of these specific regulatory sequences are located just upstream of the promoter (the so-called regulator regions or upstream promoter elements), whereas others, called enhancers, may be thousands of base pairs away from the promoter and may even be downstream of the gene. Negative regulation is also accomplished through the binding of repressor proteins to silencer sequences in the DNA.

    Step 3: Posttranscriptional mRNA Processing. Any mechanism that interferes with the formation of mature mRNA from the primary transcript or blocks the transport of mature mRNA to the cytoplasm acts at this step. For example, regulatory proteins that change the ability of the spliceosome to recognize particular intron/exon splice sites are responsible for the wide variety of antibodies in the immune system. Regulatory proteins that block the necessary modification of the 5′ cap structure, thereby preventing the mRNA from being translated, have also been identified.

    Step 4: Protein Synthesis. Translational regulation occurs most often by affecting the ability of the ribosome to initiate protein synthesis. Although this method of regulation is not as common as regulation at the transcriptional level, many translational regulation mechanisms nevertheless exist. For example, some regulatory proteins cause the formation of unusual structures in the mRNA template that prevent binding of the ribosome. tRNAs can also be selectively degraded, making certain amino acids unavailable for protein synthesis.

Step 5: Posttranslational Processing. Many newly translated polypeptides need to be posttranslationally modified to form a functional protein. For example, some proteins have to be cleaved, whereas others require chemical modifications such as the addition of sugar or phosphate groups, which either participate in the active sites of enzymes or change the conformation of the protein to the final active conformation. Regulation can occur at any one of the steps involved in modifying a protein. Some proteins are also modified by the addition of a "destabilizing" N-terminal residue that becomes ubiquitinylated by the covalent linkage of molecules of the protein ubiquitin. The ubiquitinylated protein is then digested by proteases.

Check Your Performance

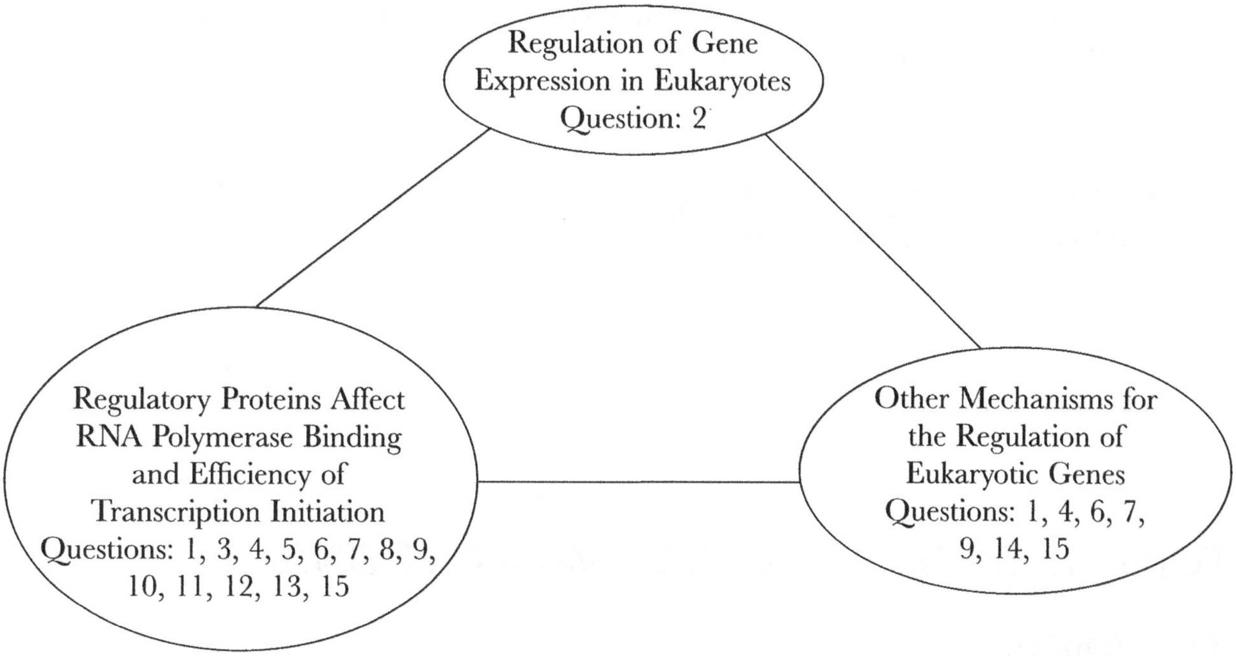

Note the number of questions in each grouping that you got wrong on the chapter test. Identify where you need further review and go back to relevant parts of this chapter.

CHAPTER 7

Recombinant DNA

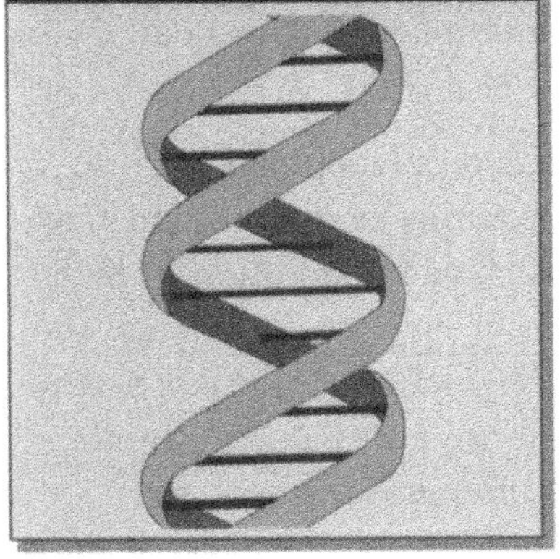

Naturally occurring DNA molecules are very long and contain many genes. Even the chromosome of *Escherichia coli*, which is approximately 1,000-fold smaller than the human genome, contains 4.6 million nucleotides and approximately 4,000 genes. Moreover, as mentioned in Chapter 6, in most eukaryotes the genes themselves account for only a small fraction of the chromosomal DNA, the rest being comprised of noncoding DNA sequences that occur between genes as well as within genes (introns). A human gene, for example, might constitute only 1/100,000 of the DNA molecule on which it resides. The huge size of chromosomes, coupled with the fact that DNA is chemically featureless, with no landmarks to distinguish a gene sequence from noncoding DNA, made it virtually impossible, before the 1970s, for scientists to study specific genes.

ESSENTIAL BACKGROUND

- **Structure of bacterial and eukaryotic DNA**
- **RNA processing in eukaryotes**

TOPIC 1: GENERATING RECOMBINANT DNA

KEY POINTS

✓ *What discoveries made it possible for scientists to isolate specific genes from large genomic DNAs?*

✓ *What is meant by "recombinant DNA" and how is it made?*

✓ *How are foreign genes transferred into cells?*

✓ *How can functional eukaryotic proteins be produced in bacterial cells if most bacteria lack the enzymes needed to remove intron sequences?*

The problem of how to isolate specific genes was resolved with the discovery of two naturally occurring classes of biological molecules—**restriction endonucleases** (or more simply restriction enzymes) and **plasmids.** Together, these two molecules made it possible to isolate, manipulate, and amplify specific pieces of DNA using techniques broadly referred to as "**genetic engineering**" or "**recombinant DNA**" technologies.

RESTRICTION ENZYMES CUT DNA AT SPECIFIC SEQUENCES

Restriction enzymes were discovered in bacteria in the late 1960s when scientists were investigating how certain bacteria protect themselves against invading viral DNA. They discovered that some bacteria produce enzymes (restriction enzymes) that can recognize foreign DNAs and degrade them. Most restriction enzymes are highly specific, recognizing short (usually four to eight nucleotides) sequences and cutting at specific points within these sequences. The bacterium protects its own DNA from restriction by adding methyl (—CH_3) groups to bases within the sequences recognized by the restriction enzyme. Because the recognition sequences for restriction enzymes are short, they will occur purely by chance in any long DNA molecule and because they are specific, a given restriction enzyme will always cut a particular DNA at the same sites, making it possible to isolate predictable fragments from any given DNA. The sequences recognized by most restriction enzymes are palindromes, in that the two DNA strands have the same sequence of four to eight nucleotides but running in opposite directions (**Figure 7.1A**), and most of these sequences are cut in one of two distinct ways. Thus, some restriction enzymes cut through the center of their recognition sequences yielding DNA fragments with "**blunt**" (double-stranded) ends. However, for most restriction enzymes, the cut sites within the recognition sequence are not directly opposite one another but are offset. The result of such an asymmetric cleavage of the DNA is that after restriction, each end of the cut DNA terminates in a short stretch of single-stranded nucleotides. These single-stranded overhangs are referred to as "**sticky**" (or **cohesive**) ends because they are complementary to one another and can stick or base pair back together again (Figure 7.1A and 7.1B). It follows that the fragments obtained from the DNA of one organism will have the same sticky ends as the fragments produced by cutting the DNA from another organism with the same restriction enzyme. This complementarity between the ends of fragments produced by cutting DNAs from different sources with the same restriction enzyme is one of the cornerstones of recombinant DNA technology. Thus, because of their complementarity, the sticky ends of DNA fragments will form hydrogen-bonded base pairs with the sticky ends of any other DNA fragments produced by the same restriction enzyme. The unions formed in this way are temporary but can be made permanent by the enzyme **DNA ligase** that seals the base-paired ends by catalyzing the formation of phosphodiester bonds. The resulting hybrid DNA fragment consisting of DNA from two different sources is referred to as **recombinant DNA** and could involve fragments from any two organisms (Figure 7.1B).

VECTORS ARE USED TO CLONE DNA FRAGMENTS

In practice, the fragments that are joined together to make recombinant DNA are not usually random pieces of genomic DNA; instead, one of the pieces of DNA usually carries a gene from a eukaryote or bacterium and the other piece of DNA is usually a **vector molecule** that acts as a vehicle to transfer the gene of interest into a bacterial or eukaryotic host cell. Two types of vectors are commonly used—one type is derived from viruses that infect bacterial cells (bacteriophages or phages) and the other type is derived from bacterial plasmids. A bacteriophage acts like a hypodermic syringe, injecting its DNA (plus any foreign DNA that has been inserted into it) into the bacterial host cell where it replicates to form many copies of itself. Bacteriophages lambda and M13 are commonly used vectors and have the advantage of being able to accept large pieces (up to 40 and 10 kb, respectively) of foreign DNA. More recently, some animal viruses have also been used to introduce foreign genes into animal and plant cells. However, the

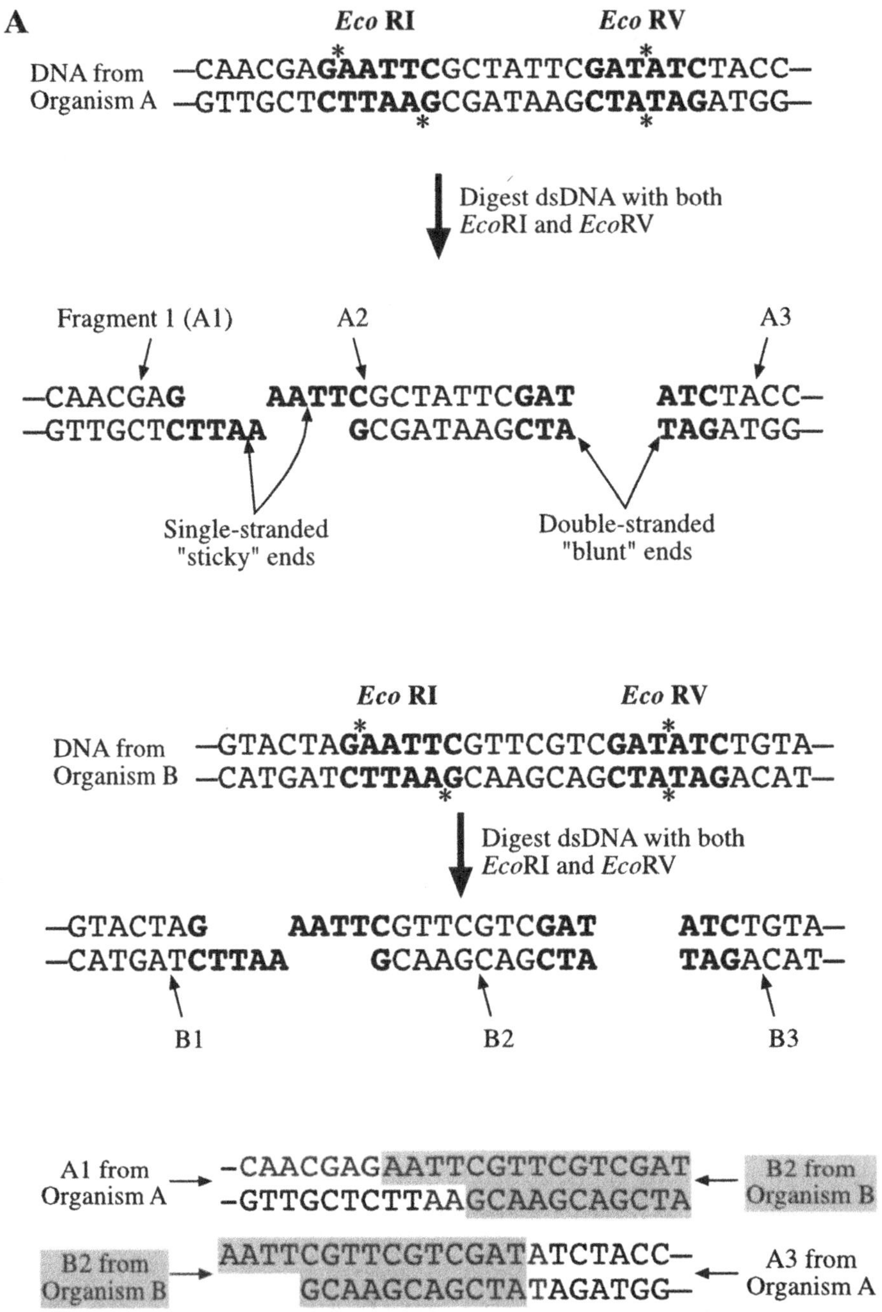

Figure 7.1 A. Restriction enzyme digestion. The DNA from organism A contains two palindromic sequences, one recognized by the restriction enzyme *Eco*RI and one by *Eco*RV. Asterisks indicate the sites at which the enzymes cut the DNA. When treated with both enzymes, three fragments result, labeled A1, A2, and A3. *Eco*RI digestion of organism A's DNA generates fragments with "sticky" ends, so-called because the single-stranded ends produced will form hydrogen-bonded base pairs with the sticky ends of any other DNA fragments produced by the same restriction enzyme. *Eco*RV produces fragments with "blunt" ends, so-called because no single-stranded DNA is formed. B. Generation of recombinant DNA. Organism B contains sites for *Eco*RI and *Eco*RV, and when digested with both enzymes, three fragments result with ends similar to those from Organism A. C. Recombinant DNA is produced when a DNA fragment from one organism is joined with a DNA fragment from a different organism. Because the "sticky" ends produced by *Eco*RI digestion of DNA from both organism A and organism B are complementary, fragments A1 and B2 can base pair together. Because no base pairing is involved, any two "blunt"-ended fragments can be joined together.

most commonly used vectors for transferring foreign DNA into cells are bacterial plasmids, which accept smaller pieces (5–10 kb) of foreign DNA, and the rest of this discussion on vector molecules focuses on them.

A typical plasmid vector is a relatively small circular DNA molecule that contains its own origin of replication (*ori*) so it can replicate inside a host cell independent of the host chromosome. Although plasmids occur naturally in bacteria, where they usually confer useful properties on the cell such as antibiotic resistance, most plasmid vectors in use today have been specially modified or constructed for the purpose of transferring foreign DNA into cells. Some, such as pUC18 and the Bluescript^{R} vectors, have been engineered to make many hundreds of copies of themselves, and thus any foreign DNA they contain, once they are introduced into a host cell. This process of producing a large quantity of identical copies of any chosen DNA is referred to as **cloning**, and the population of cells derived from a host cell containing a recombinant plasmid (or recombinant phage DNA) is referred to as a **cell clone** as each new cell formed will also contain multiple copies of the recombinant vector.

For cloning to work, the vector must have a cutting site for a restriction enzyme that makes it possible to insert the desired fragment of DNA and, second, the vector must be selectable so that cells that have taken up a vector can be identified. To overcome the difficulty of inserting fragments into vectors, most plasmids have been constructed with **polylinkers** (also called **multiple cloning sites**)—pieces of DNA that contain one or more unique restriction enzyme recognition sites. Moreover, most plasmids contain **antibiotic-resistance genes** or other **selectable markers** that enable cells containing plasmids to be isolated from cells that have not taken up a plasmid.

GENOMIC AND COMPLEMENTARY DNA LIBRARIES ACT AS SOURCES OF DNA FOR CLONING

Cloning is used for two general purposes. Thus, the goal may be to produce the protein encoded by the foreign gene. For example, all human insulin is now made in large quantities from recombinant plasmids in bacteria. Alternatively, the goal may be to produce many identical copies of any given DNA fragment so that the DNA can be further characterized. However, regardless of the purpose of the cloning, most cloning procedures do not start with a single isolated eukaryotic or bacterial gene fragment but instead start with the total genomic DNA of an organism. Because even a typical bacterial genome is cut into several hundred to several thousand fragments with most of the commonly used restriction enzymes, thousands of different recombinant plasmids are actually obtained from most standard cloning procedures (**Figure 7.2**). The mixture of recombinant plasmids is then introduced by a process referred to as **transformation**, into a suitable host, such as *E. coli*. (For cloning purposes, special modified strains of *E. coli* are used that are defective in their restriction activity [denoted as r^- strains]. Such r^- strains cannot recognize foreign DNA and so do not degrade the plasmid DNA.) The resulting population of cells, each carrying a recombinant plasmid with a particular segment of a total genome, is referred to as a **genomic library**. Because many recombinant plasmids will contain identical fragments, to assemble a complete library of the entire source genome, several hundred to several million clones could be required. For example, a complete library of the human genome consisting of fragments 20 kb long would require close to a million clones to cover the entire genome.

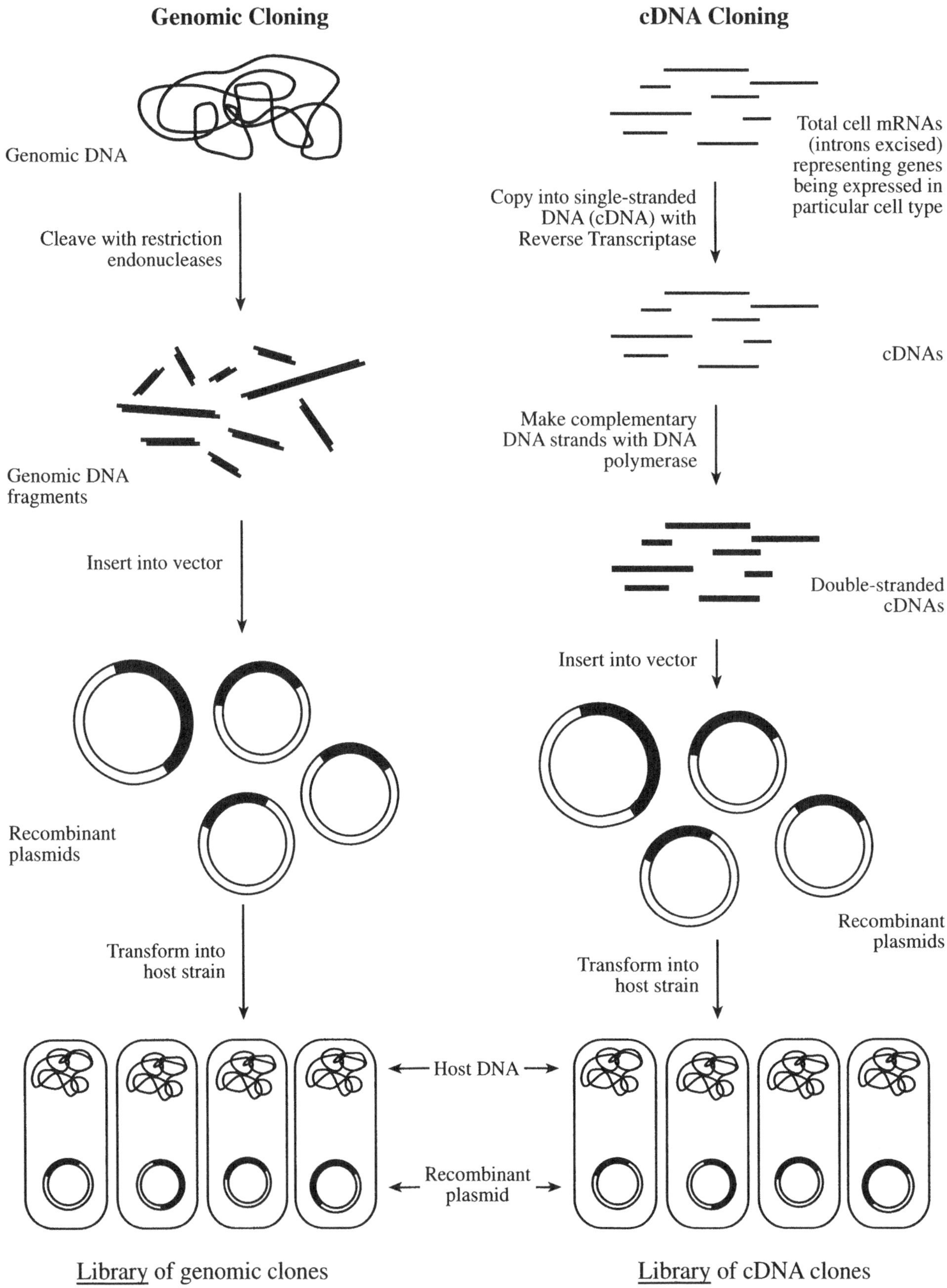

Figure 7.2 Construction of genomic and cDNA libraries.

Because a genomic library is prepared from total chromosomal DNA, the genomic library of a eukaryote would include not only coding regions but also introns, noncoding sequences that interrupt most eukaryotic genes. If the goal of the cloning is to produce a eukaryotic protein in a bacterial cell, a genomic library cannot be used as the source of the eukaryotic gene because bacteria do not possess RNA-splicing machinery and so cannot remove introns. The difficulty of obtaining "expressible" eukaryotic genes was overcome with the discovery of a viral enzyme called **reverse transcriptase** that can copy RNA into DNA. Thus, by isolating **mRNAs** from eukaryotic cells and using reverse transcriptase to make DNA copies of these mRNAs, scientists can now obtain copies of eukaryotic genes without the introns (Figure 7.2). This DNA is called **complementary DNA** (**cDNA**) and, like genomic DNA fragments, cDNA can be attached to vector DNA for replication inside a host cell. However, because mature mRNA contains none of the regulatory sequences that precede genes (e.g., promoter sequences), specialized "**expression" vectors** are used for cloning eukaryotic cDNAs. These expression vectors carry bacterial promoter sequences immediately adjacent to the polylinker region where the cDNAs are inserted, thereby providing the bacterial RNA polymerase with the sequences necessary for transcription of the eukaryotic cDNA gene. It is also important to realize that a **cDNA library** represents only part of the genome of a cell: only those protein-coding genes that were transcribed in the cells from which the mRNAs were isolated.

By using genes isolated from cDNA libraries, many functional eukaryotic proteins can be produced in bacterial cells. However, some eukaryotic proteins cannot be produced using bacteria as the host. For example, some eukaryotic proteins are lethal to bacterial cells, whereas others can be made but in a nonfunctional form because they need to be modified after translation (often by the addition of phosphate or sugar residues) and the bacterial host cells cannot perform these specific posttranslational processing functions. In these cases, a eukaryotic host, normally yeast, is used. Yeast cells are a rarity among eukaryotes in that, like bacteria, they contain plasmids and they are easy to grow. However, because they are eukaryotic cells, they can carry out many posttranslational processing functions that bacterial host cells cannot perform and other eukaryote-bacteria incompatibilities are also avoided.

POLYMERASE CHAIN REACTION CAN AMPLIFY DNA IN THE TEST TUBE

Cloning (the amplification of a piece of foreign DNA by inserting it into a vector molecule) is not the only way to obtain many copies of a particular DNA fragment for analysis. In fact, if part of the sequence of the desired DNA fragment is known, large quantities of this DNA can be obtained much more rapidly by a technique known as the **polymerase chain reaction**, or **PCR**. PCR is based on multiple rounds of DNA replication in the presence of **Taq DNA polymerase** (a heat-stable DNA polymerase isolated from the thermophilic bacterium, *Thermus aquaticus*). The desired DNA is flanked by two different synthetic primers that are complementary to sequences on opposite strands at opposite ends of the double-stranded DNA fragment to be amplified. Because these primers have to be chemically synthesized, this explains why PCR can only be used to specifically amplify a piece of DNA whose beginning and end sequences are known. PCR starts by denaturing the double-stranded target DNA to give single strands that can serve as templates for DNA replication. After cooling, the primers hybridize to the 3′ ends of each of these template strands, and Taq DNA polymerase then catalyzes the synthesis of new complementary strands, effectively doubling the amount of double-stranded target DNA. The cycle—denaturation, primer annealing, DNA synthesis—is repeated many times, each cycle doubling the number of double-stranded target DNA molecules and hence the number of single-stranded template (**Figure 7.3**). Thus, if the original reaction contained one copy of the target DNA, after 10 cycles, there would be more than a 1000 copies, and after 20 cycles, more than one million copies. PCR is an extremely useful technique because it allows the rapid amplification of DNA sequences from minute quantities of starting material. For example, PCR was successfully used to amplify recognizable sequences from the few DNA molecules remaining in the 18-million-year-old fossilized remains of an ancient magnolia plant. PCR is also routinely used in forensics to amplify DNA from tiny amounts of blood, tissue, or semen found at a crime scene. Furthermore, the specificity of the PCR primers means that the target DNA does not have to be purified: PCR can be used to amplify sequences directly from a cell lysate.

Topic Test 1: Generating Recombinant DNA

True/False

1. If double stranded, the DNA sequences 5′-AATT-3′ and 3′-AATT-5′ would be cut by the same restriction enzyme.

2. Pharmaceutical companies that use bacteria carrying human genes to produce large quantities of human proteins, such as insulin, use genomic libraries as a source of the human genes.

Short Answer

3. PCR is based on the repetition of cycles of DNA replication. Each cycle involves three basic steps. What are these steps, and in what order do they occur?

4. How many potential 6 base pair restriction enzyme recognition sites are there in the following segment of double-stranded DNA?

5′-TAGACGTCGAAGGATCCAAAGGG-3′
3′-ATCTGCAGCTTCCTAGGTTTCCC-5′

5. The restriction enzyme *Bgl*II recognizes the sequence 5′-AGATCT-3′ and cleaves on the 5′ side to the left of the G. (Because the top and bottom strands of most restriction sites read the same in the 5′-to-3′ direction, only one strand of the site is shown.) The single-stranded ends produced by *Bgl*II cleavage are identical to those produced by *Bam*HI cleavage (as shown below), allowing the two types of ends to be joined together upon incubation with DNA ligase.

5′-GGATCC-3′	*Bam*HI →	–G	GATCC–
3′-CCTAGG-5′		–CCTAG	G–
5′-AGATCT-3′	*Bgl*II →	–A	GATCT–
3′-TCTAGA-5′		–TCTAG	A–

If a fragment with a *Bgl*II sticky end is ligated to a fragment with a *Bam*HI sticky end, could the resulting *Bgl*II/*Bam*HI hybrid site be cut with *Bgl*II?

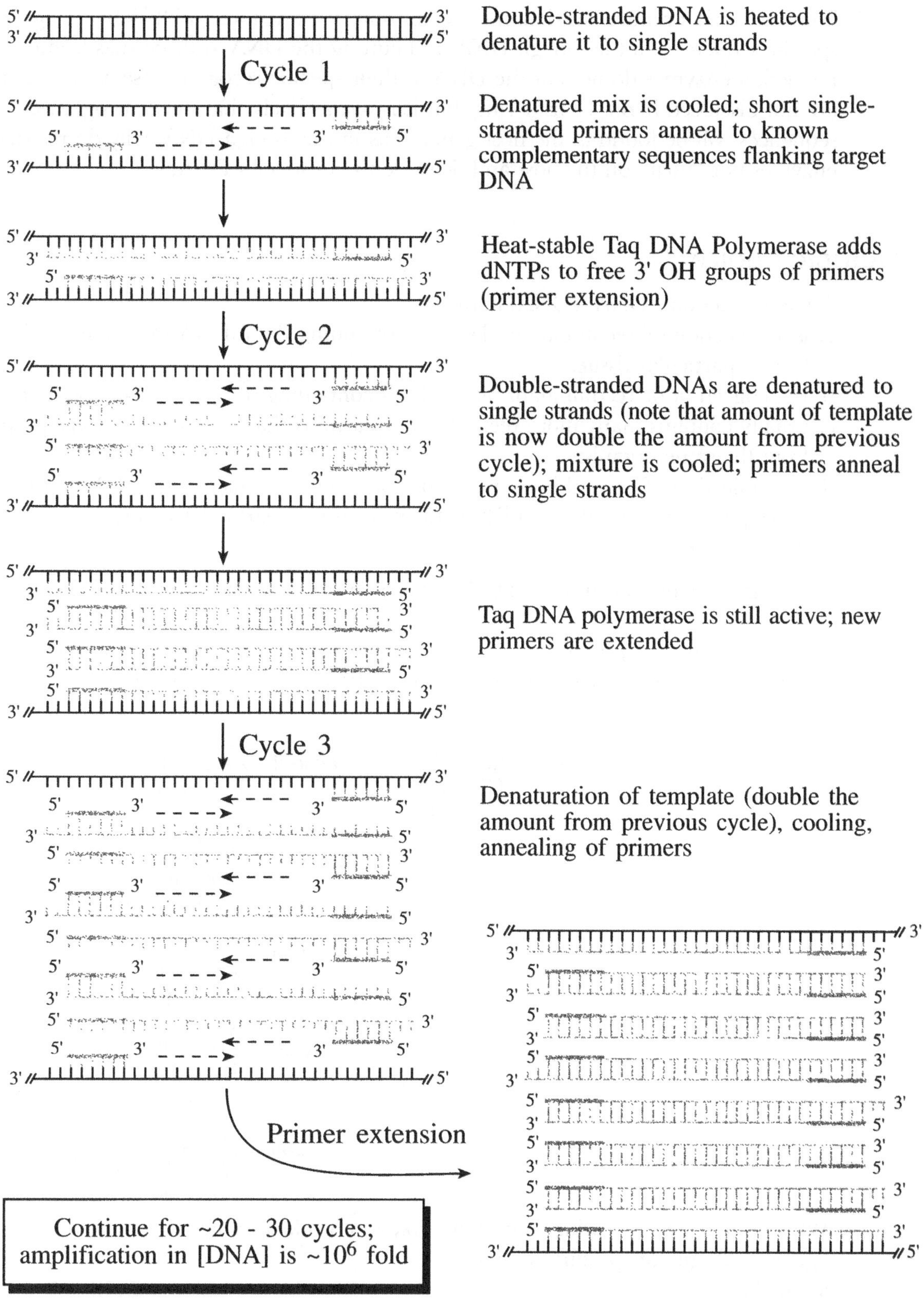

Figure 7.3 Polymerase chain reaction.

6. Some bacteria produce restriction enzymes that degrade foreign DNA by recognizing specific sequences in the foreign DNA and cutting the DNA at those sites (some restriction enzymes do not cut the DNA at their specific recognition sequences but they are not discussed here). Considering that it is just as likely that those recognition sequences will be found in the host genome as in the foreign DNA, why do restriction enzymes not "turn" on the host and degrade the host cell's DNA?

Multiple Choice

7. A complementary DNA (cDNA) library
   a. is a collection of recombinant plasmids containing the mRNA molecules purified from a particular tissue.
   b. is a collection of recombinant plasmids all containing the same DNA fragment.
   c. usually contains many more recombinant plasmids than a genomic library prepared from the same organism.
   d. is a collection of recombinant plasmids containing complementary DNA inserts corresponding to all of the mRNAs purified from a particular tissue.
   e. None of the above

8. The fragments produced when DNA is cut with the restriction enzyme *Spe*I can be *directly* (where "directly" means without any modification of the ends of the fragments) ligated to fragments produced by one of the restriction enzymes listed on the right below. (The arrows indicate where each restriction enzyme cuts within its recognition sequence.)

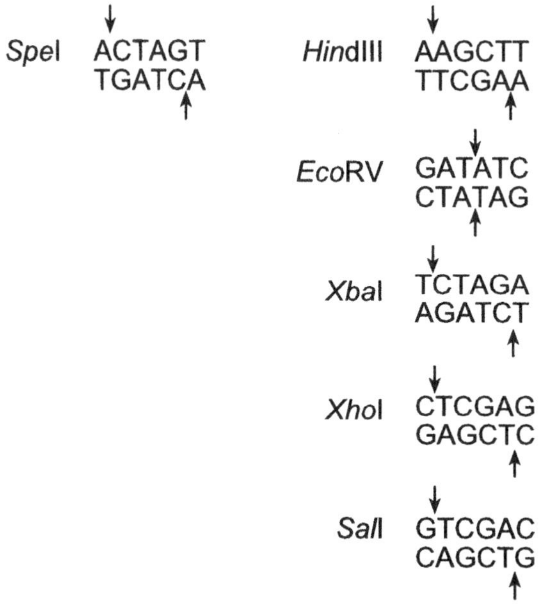

Which one of the following restriction enzymes produces fragments that can be directly ligated to fragments produced by *Spe*I?
   a. *Hind*III
   b. *Eco*RV
   c. *Xba*I
   d. *Xho*I
   e. *Sal*I

9. To be able to amplify a piece of DNA using the polymerase chain reaction, it is necessary to
   a. have large quantities of DNA to start with.
   b. have pure DNA to start with.
   c. know a short sequence flanking the 3′ end of each strand of the piece of double-stranded DNA to be amplified.
   d. know a short sequence flanking the 5′ end of each strand of the piece of double-stranded DNA to be amplified.
   e. All of the above

10. Pick the pair of primers that you could use to amplify the following double-stranded DNA sequence using PCR:

    5′-CCACTAGGGAGCTACGTAGGCACGGCATTACACGGATAGGCATTAACG-3′
    3′-GGTGATCCCTCGATGCATCCGTGCCGTAATGTGCCTATCCGTAATTGC-5′

   a. 5′-GGTGATC-3′; 5′-GCAATTA-3′.
   b. 5′-GGTGATC-3′; 5′-CGTTAAT-3′.
   c. 5′-CCACTAG-3′; 5′-GCAATTA-3′.
   d. 5′-CCACTAG-3′; 5′-CGTTAAT-3′.
   e. 5′-GATCACC-3′; 5′-CGTTAAT-3′.

Topic Test 1: Answers

1. **False.** When written out in double-stranded form and in the same orientation

   5′-AATT-3′ 5′-TTAA-3′
   3′-TTAA-5′ 3′-AATT-5′

   it is obvious that the sequences are different, despite the fact that they are both palindromes. Most restriction enzymes recognize a single specific DNA sequence so it is very unlikely that these two sequences would be cut by the same restriction enzyme.

2. **False.** A genomic library is a collection of recombinant plasmids containing all of the coding and noncoding sequences of a particular organism. Such libraries, if made from eukaryotic genomes, would contain introns. Therefore, if the purpose of the cloning is to produce a functional eukaryotic protein inside a bacterial host cell, a genomic library of the eukaryote in question could not be used as the source of the eukaryotic gene because bacteria are incapable of removing introns. To obtain a copy of the eukaryotic gene that can be expressed and translated into a functional protein in bacteria, the eukaryotic gene would have to be isolated from a complementary DNA (cDNA) library.

3. **Template denaturation, primer annealing, and DNA replication (or primer extension).**

4. 5′-GACGTC-3′ and 5′-GGATCC-3′
   3′-CTGCAG-5′ 3′-CCTAGG-5′
   are the only two palindromic sequences in this stretch of DNA.

5. **No.** If a fragment with a *Bgl*II sticky end was ligated to a fragment with a *Bam*HI sticky end, either one of the following double-stranded sequences could be generated:

—GGATCT— —AGATCC—
—CCTAGA— —TCTAGG—

These double-stranded sequences are not palindromes, and neither one matches the recognition sequences for either *Bam*HI or *Bgl*II. Therefore, neither *Bgl*II nor *Bam*HI would be able to cut the new sequences generated by ligating a *Bgl*II end to a *Bam*HI end.

6. Methylation of bases in the recognition sequence for the host's restriction enzyme(s) prevents the host DNA from being degraded by its own enzyme(s).

7. **d.** Only double-stranded DNA can be cloned so a is incorrect. cDNA is synthesized from all the messenger RNAs (mRNAs) isolated from a particular tissue or cell type. Because the mRNAs present in a particular tissue or cell type will represent all of the protein-coding genes expressed in that cell type, many different DNAs will be made and cloned, so b is not correct either. Finally, a cDNA library does not contain DNAs corresponding to ribosomal RNA or transfer RNA genes or to the huge number of noncoding sequences present in most eukaryotic genomes or to all the other protein-coding genes that are not expressed in that particular cell type. Therefore, a cDNA library will be considerably smaller than a genomic library, which represents all the coding and noncoding sequences of a particular organism. This eliminates c, leaving d, which is a correct statement.

8. **c.** When the DNA is cut with *Spe*I, "CTAG" and "GATC" sticky ends are produced. For these fragments to be directly ligated with fragments resulting from digestion with another enzyme, that other enzyme must produce the same sticky ends. Only *Xba*I fits this description, also producing "CTAG" and "GATC" sticky ends.

9. **c.** PCR's usefulness lies in its ability to amplify a specific DNA sequence from a very small amount of impure starting material. For this reason, a and b are incorrect. To begin DNA synthesis, a primer is necessary to provide a free 3′ OH group for the DNA polymerase. Because nucleotides can only add on to the 3′ end of a growing DNA or RNA chain and because the two DNA strands in a double helix are antiparallel (i.e., they run in opposite directions), it follows that the PCR primers must be able to hybridize to the 3′ end of each of the template strands in order for synthesis of complementary DNA to occur in the correct direction as illustrated below:

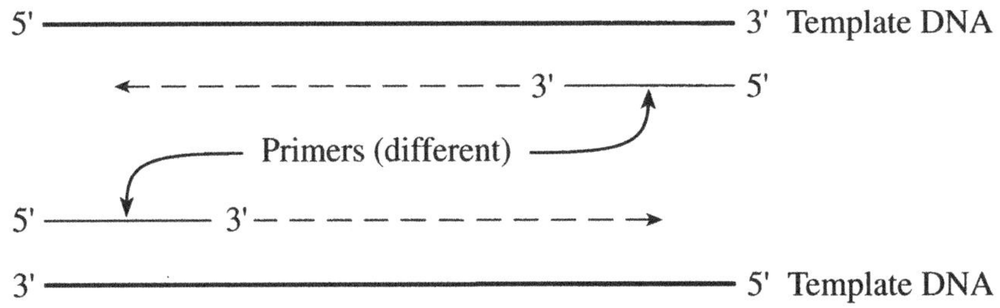

Therefore, it is necessary to know some of the sequence flanking the 3′ end of each of the template strands to make PCR primers that will hybridize to them and "prime" DNA synthesis. This is described in c, which is therefore the correct option.

10. **d.** This is the only pair of primers that are complementary to sequences at the 3′ end of each of the template strands of the target DNA sequence. See the answer to question 9

above for an explanation of why the PCR primers need to be complementary to the 3′ ends of the target DNA sequence.

TOPIC 2: ISOLATING SPECIFIC RECOMBINANT CLONES

KEY POINTS

- ✓ *How can specific recombinant clones be isolated from the several hundred to several million clones present in a typical genomic or cDNA library?*
- ✓ *How can pieces of DNA that will hybridize to sequences in a gene of interest be made if the nucleotide sequence of the desired gene is not known?*

The process of generating recombinant DNA can be divided into four steps, three of which (the isolation of the desired DNA fragments, the insertion of these fragments into a suitable vector such as a plasmid, and the introduction of the vector into an appropriate host) have already been described. The fourth step involves the identification of a recombinant clone carrying a fragment or gene of interest and the isolation and characterization of this clone.

"MARKER" GENES ON THE PLASMID IDENTIFY HOST CELLS THAT CARRY RECOMBINANT DNA

As explained in the previous section, in most cloning strategies, a foreign genome is digested with a restriction enzyme and the resulting fragments ligated to a cut vector to give thousands of different recombinant vector molecules, each containing a different piece of the foreign genome. To make the isolation of a clone carrying a particular piece of the genome easier, vectors are chosen that make it easy to eliminate from the library any cells that do not contain vector molecules and any cells with vectors that do not contain foreign DNA (i.e., cells containing re-annealed vectors). For example, both of these types of cells can be eliminated by using a plasmid vector that contains two or more antibiotic resistance genes. The foreign genome and the plasmid are digested with a restriction enzyme that cuts within one of the two antibiotic resistance genes. After the genomic and plasmid DNAs are ligated together and transformed into a host cell, the transformants are plated onto medium that contains the antibiotic whose resistance gene is not cut by the restriction enzyme. Only those host cells that contain plasmids will survive and form colonies (cell clones), regardless of whether the plasmids contain foreign DNA or not. To distinguish between cells that carry plasmids with foreign DNA inserts and those that contain plasmids that have recircularized without foreign DNA, cells from the surviving colonies are then subcultured (**replica plated**) onto medium containing the antibiotic whose resistance gene is interrupted by the insertion of the foreign DNA. The cells that fail to grow on this medium should contain plasmids that carry foreign DNA sequences and these sensitive clones can be retrieved from the first selection plates (**Figure 7.4**).

In addition to genes for antibiotic resistance, several other marker or "reporter" genes that, unlike the antibiotic resistance genes, do not kill cells lacking them are used to isolate the desired recombinant product. For example, several plasmids have been constructed that include unique sites for restriction enzymes within the *lacZ* gene, which encodes β-galactosidase (see Chapter 5 for a discussion of the *lac* genes). Insertion into any one of these restriction enzyme recognition sites disrupts expression of the *lacZ* gene and leads to the loss of β-galactosidase activity, which is

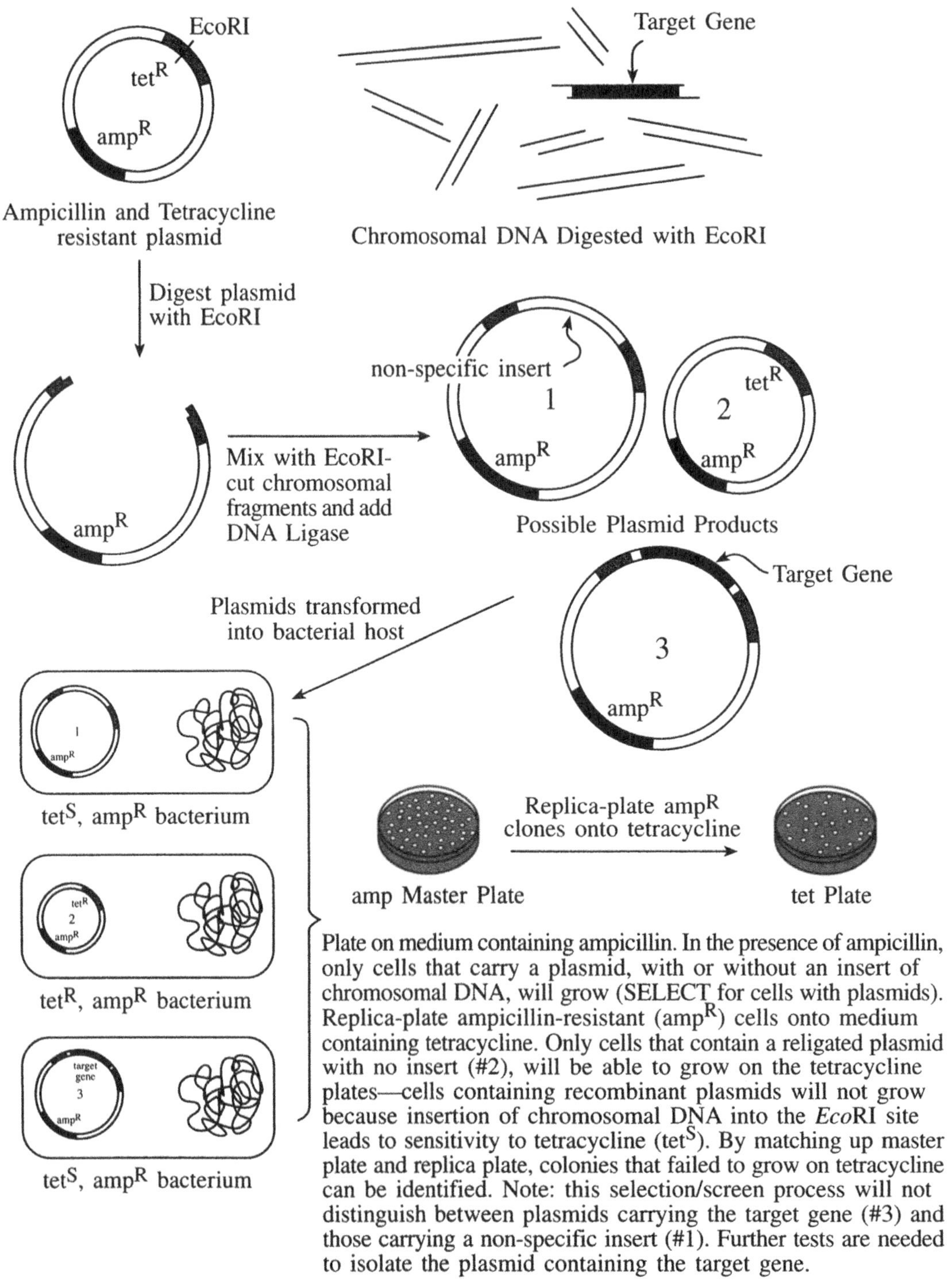

Figure 7.4 Using antibiotic-resistance genes to isolate recombinant clones.

readily monitored in the host cells by growing the cells in the presence of X-Gal. X-Gal is a substrate of β-galactosidase and is converted by the enzyme into a blue-colored product. Therefore, in the presence of X-Gal, cells capable of producing β-galactosidase turn blue, whereas cells lacking β-galactosidase activity remain white. Thus, by growing the transformants in the presence of X-Gal, recombinant cells that carry a copy of the plasmid with an insert of foreign DNA can be easily identified as those that form white colonies (**Figure 7.5**). Newer "reporter"

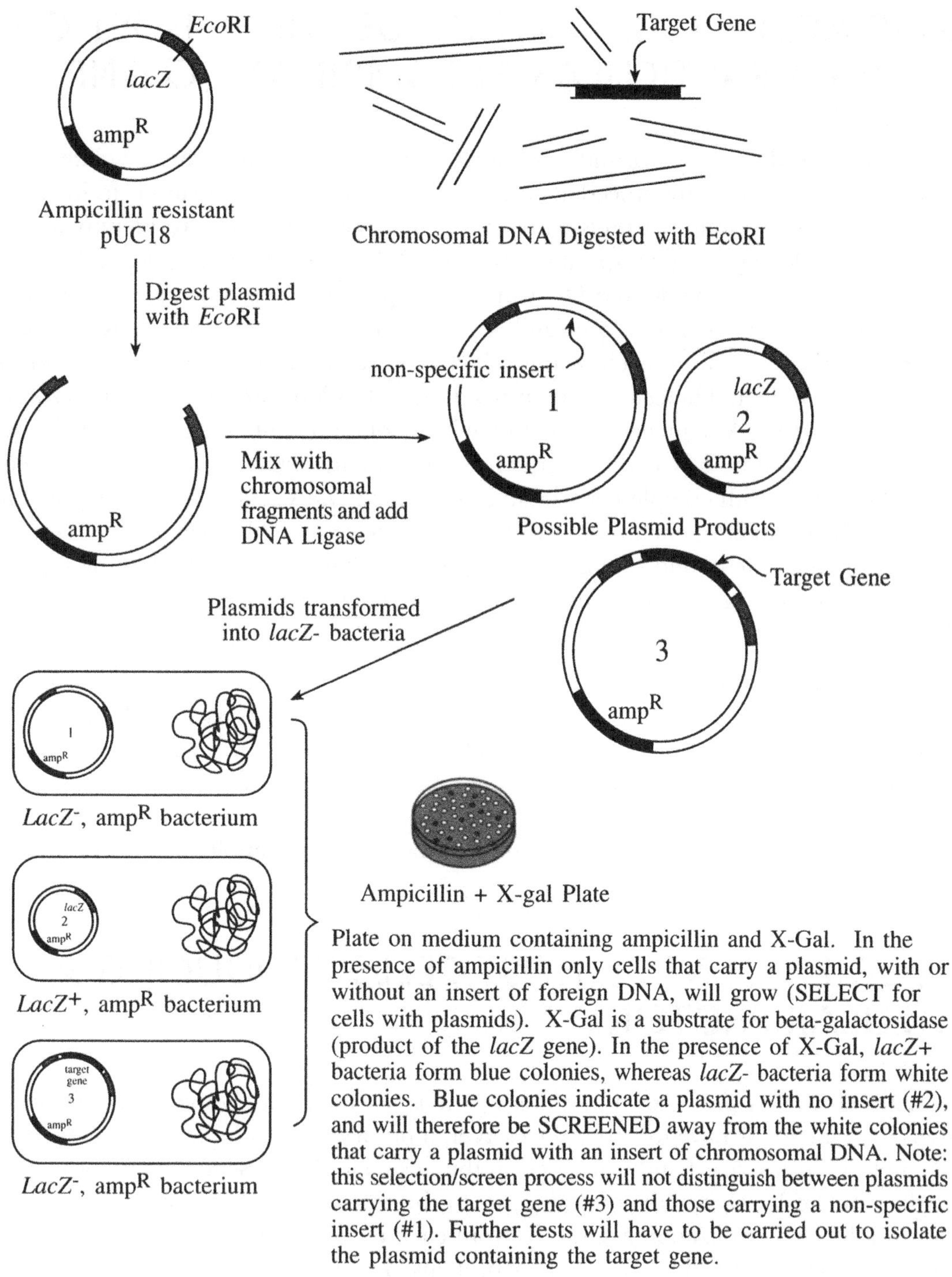

Figure 7.5 Isolation of recombinant clones using reporter genes on the vector.

genes include the gene for firefly luciferase and the gene for green fluorescent protein both of which cause cells to fluoresce under the appropriate conditions. Because testing for the loss of β-galactosidase activity or fluorescence does not kill the cells lacking these activities, when reporter genes are used to identify host cells carrying recombinant clones, this step is referred to as a "**screen.**"

NUCLEIC ACID HYBRIDIZATION IDENTIFIES HOST CELLS CARRYING A PARTICULAR RECOMBINANT CLONE

Once host cells that carry plasmids with a foreign DNA insert have been isolated using selection and/or screening procedures as described above, the next step is to identify those clones that contain plasmids with a particular piece of DNA. To identify a recombinant clone carrying a specific piece of DNA, a technique that depends on base pairing between the desired DNA and a complementary sequence on another nucleic acid molecule called a "**probe**" is used. In this approach, called **nucleic acid hybridization**, cells from each of the cell clones are first transferred onto a solid support, usually a nylon filter. The cells are then lysed, their DNA denatured, and the resulting single-stranded DNA sequences incubated with a solution containing a radioactively labeled single-stranded probe DNA whose sequence is complementary to part of the desired DNA. The probe hybridizes with complementary single-stranded DNA sequences on the filter, and after washing, the location of the bound probe DNA is determined by placing the filter in contact with x-ray film. The bound radioactivity exposes the film, and by aligning the exposed areas on the film with the original colony plates, clones carrying the desired fragment can be identified (**Figure 7.6**).

Obviously, for this procedure to work, at least part of the sequence of the gene of interest must be known to be able to construct the probe. This information is usually obtained by isolating the protein encoded by the gene of interest and using the genetic code to work backward from the amino acid sequence of a small part of this protein. Because of the degeneracy of the genetic code (explained in Chapter 3), several probes may need to be constructed or else the conditions arranged such that the probe does not need to be perfectly complementary to the desired sequence to hybridize to it.

ANTIBODIES IDENTIFY HOST CELLS PRODUCING A PARTICULAR PROTEIN

Sometimes a DNA probe for a particular gene is not available. Under these circumstances, alternative methods have to be used to screen a library. For example, if the cloned DNA is able to be transcribed and translated in the host cell, the presence of the protein can be determined by an immunological assay. In this technique, which has a lot in common with the DNA hybridization assay, cells from each clone are transferred to a solid support where the cells are lysed to expose their proteins. The immobilized proteins are then incubated with a primary antibody that specifically binds to the protein encoded by the target gene. Unbound antibody is then washed away, and the immobilized proteins are then treated with a second antibody that is specific for the primary antibody. If the second antibody is radiolabeled or linked to an assayable enzyme, the location of the bound antibodies can be determined. By lining up the solid support with the bound antibodies with the original master plate, the clones to which the primary antibody bound can be identified. Such clones could contain a complete copy of the desired gene or a portion of the gene that is large enough to make a protein product capable of being recognized by the primary antibody. Further analysis is often necessary to determine which clones, if any, carry a copy of the entire gene.

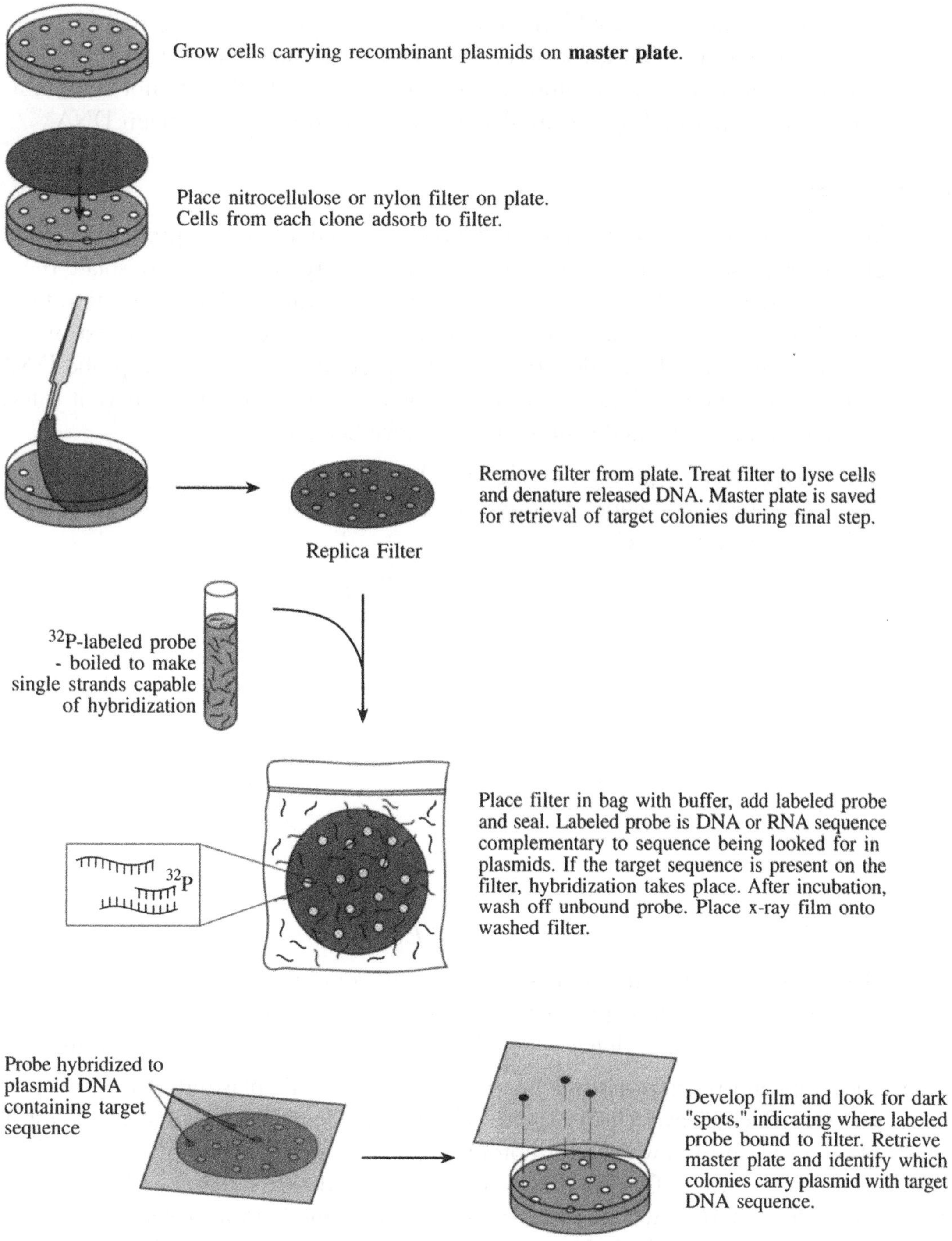

Figure 7.6 Colony screening by probe hybridization.

Topic Test 2: Isolating Specific Recombinant Clones

True/False

1. The fundamental difference between a DNA hybridization assay and an immunological assay is that in the DNA hybridization assay, a piece of DNA is used as a "probe" to locate complementary DNA sequences, whereas in an immunological assay, antibodies are used to locate and bind the target DNA sequences.

2. Plasmids containing an antibiotic resistance gene and a reporter gene with a polylinker inserted into it are often used as vectors for cloning. Such plasmid vectors make it easier to isolate cells that carry recombinant plasmids from cells that do not contain a plasmid or else contain a plasmid that recircularized without an insert of foreign DNA.

Short Answer

3. You discover a previously unreported enzymatic activity so you prepare a genomic library of the organism in which you discovered this activity and set out to isolate the relevant gene. To identify cells carrying this new gene using the DNA hybridization techniques described in the text, at least part of the sequence of the gene of interest must be known so that a probe can be made. How would you go about preparing a probe DNA for "your" new gene if you do not know any of its sequence or even where it is located in the genome and if no related gene sequences have been identified?
4. You want to clone a piece of foreign DNA into the plasmid vector shown below (where kan^R = a gene that encodes resistance to kanamycin and str^R = a gene that encodes resistance to streptomycin).

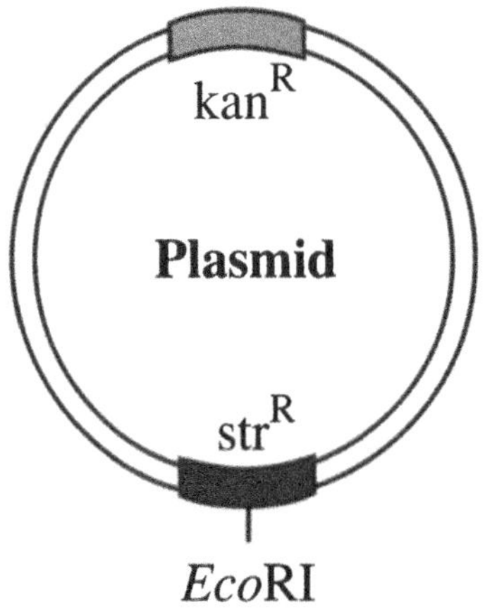

Briefly outline the sequence of steps that you would follow to
a. insert a piece of foreign DNA into this plasmid vector, and
b. isolate bacteria carrying a copy of the plasmid with an insert of foreign DNA.

5. An ampicillin-resistant tetracycline-resistant (amp^R tet^R) plasmid (plasmid 1) is treated with *Bam*HI, which cleaves the tet^R gene. The DNA is ligated with a *Bam*HI digest of another plasmid (plasmid 2) which is streptomycin-resistant (str^R) and the ligated DNAs used to transform a strain of *E. coli* that is sensitive to all three of these antibiotics. Which antibiotic or antibiotics would you add to the culture medium to select for bacteria containing plasmid 1 into which the streptomycin-resistance gene from plasmid 2 has been inserted?

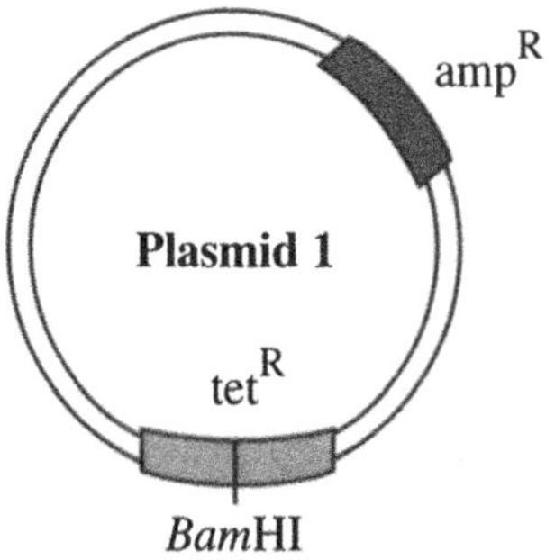

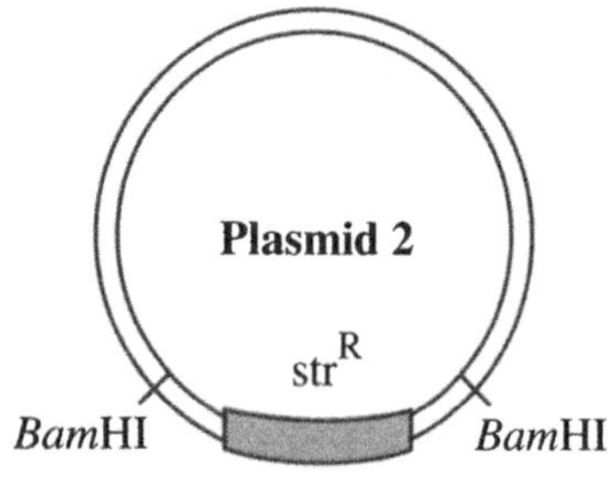

Multiple Choice

6. The following two plasmids were cut with *Eco*RI, the resulting fragments mixed together, and DNA ligase added.

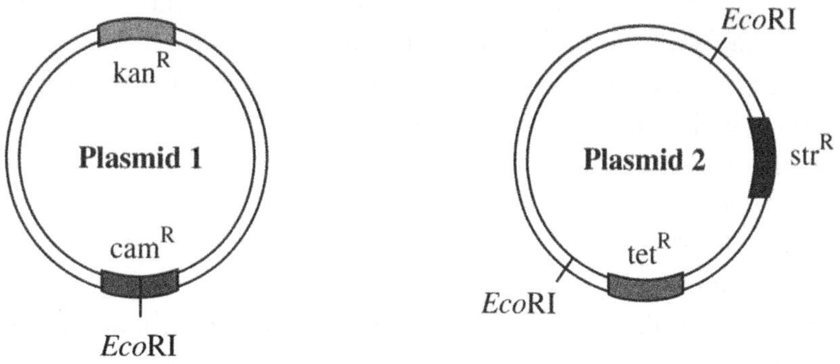

(The superscript R means that the gene confers resistance to that particular antibiotic, where kan = kanamycin, cam = chloramphenicol, tet = tetracycline, and str = streptomycin. If an antibiotic resistance gene is not shown in a plasmid, assume that the plasmid is sensitive to that particular antibiotic. The lines intercepting the plasmid indicate where the adjoining restriction enzyme cuts the plasmid.)

Using the restriction data provided in the figure, determine which one of the following antibiotic resistance phenotypes could *not* be represented among bacteria transformed with the resulting ligation products, assuming that each bacterial cell takes up only one of the ligation products. (The superscript S means that a bacterium is sensitive to that particular antibiotic.)

a. kan^R cam^S tet^S str^S
b. kan^R cam^S tet^R str^R
c. kan^R cam^R tet^S str^S
d. kan^S cam^S tet^R str^R
e. kan^R cam^R tet^R str^R

7. The following plasmid is cut with the restriction enzyme *Bam*HI that cuts through the middle of the amp^R gene.

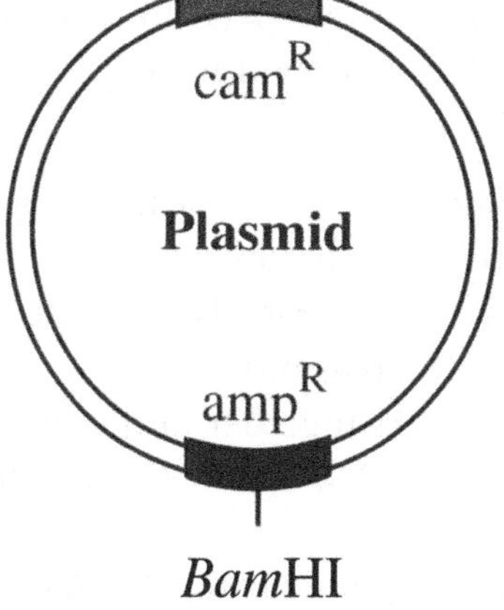

cam^R = gene encoding resistance to chloramphenicol
amp^R = gene encoding resistance to ampicillin

Genomic DNA from a mouse (which does not carry any antibiotic resistance genes) is also cut with *Bam*HI, the mouse genomic DNA fragments are mixed with the cut plasmid DNA, and DNA ligase is added. *E. coli* cells that are sensitive to ampicillin and chloramphenicol are then transformed with these ligated DNAs and grown in the *absence* of antibiotics. *E. coli* with which one of the following antibiotic resistance phenotypes could *not* result from this cloning? (Note: The superscript "S" means that the *E. coli* is "sensitive" to that particular antibiotic. Also note that the mouse DNA fragments can only enter the bacteria if they integrate into a plasmid.)

a. cam^R amp^S
b. cam^S amp^R
c. cam^R amp^R
d. cam^S amp^S
e. *E. coli* with all of the above antibiotic resistance phenotypes could result from this cloning.

Topic Test 2: Answers

1. **False.** In an immunological assay, a primary antibody that recognizes and binds to the *protein product* of the target gene, not to the target DNA itself, is used.
2. **True.** An antibiotic resistance gene on a plasmid vector allows host cells containing a copy of the plasmid (with or without an insert of foreign DNA) to be isolated from host cells that do not contain a plasmid (and hence cannot grow on medium containing the antibiotic). Reporter genes typically contain a polylinker region into which the target gene is inserted, causing loss of reporter gene activity. This loss of activity allows host cells containing plasmids with a foreign DNA insert to be *screened* away from host cells containing plasmids that recircularized without an insert of DNA.
3. To locate a particular gene sequence using the DNA hybridization technique, a small amount of the gene sequence must be known to construct a complementary probe that will be able to base pair to it. If a new gene is discovered and no sequence information is available, some sequence information can be obtained by isolating the protein encoded by the gene and using the genetic code to work backward from the amino acid sequence of a small part of this protein. Because of the degeneracy of the genetic code (explained in Chapter 3), several probes may need to be constructed or else the conditions arranged such that the probe does not need to be perfectly complementary to the desired sequence to hybridize to it.
4. To produce and isolate bacteria carrying a copy of the plasmid with an insert of foreign DNA, the steps would be as follows:
   a. Digest the plasmid and foreign DNA with *Eco*RI. *Eco*RI cuts within the streptomycin-resistance gene on the plasmid.
   b. Mix the digested DNAs together and add DNA ligase.
   c. Transform the ligated DNAs into *E. coli* cells that are sensitive to kanamycin and streptomycin.
   d. Select for resistance to kanamycin. Bacteria that do not contain a copy of the plasmid will die on medium containing kanamycin because they do not carry a copy of the gene for kanamycin resistance. Bacteria that carry a copy of the plasmid, with or without an insert of foreign DNA, will survive.

e. Test for sensitivity to streptomycin. Bacteria that carry a plasmid with an insert of foreign DNA will be sensitive to streptomycin because *Eco*RI cuts within the streptomycin-resistance gene and insertion of a piece of foreign DNA at the *Eco*RI cut site will disrupt the normal functioning of the str^R gene. Therefore, the desired cells will not grow on medium containing streptomycin, whereas bacteria that carry a copy of the plasmid without an insert of foreign DNA will grow because they will still be streptomycin resistant. Therefore to isolate bacterial clones carrying the desired recombinant plasmid, cells from each of the colonies that grow in the presence of kanamycin should be plated onto medium containing streptomycin (so-called replica plating). Cells that carry a recombinant plasmid (the desired product) will not grow on the medium containing streptomycin and they can then be retrieved from the kanamycin-resistant colonies that came up on the first selection plates.

5. **Ampicillin and streptomycin.** If the transformed *E. coli* cells were grown in the presence of tetracycline, only bacteria that contained plasmids without an insert of foreign DNA would be able to grow because an insertion into the *Bam*HI site of plasmid 1 would disrupt the normal functioning of the tet^R gene. Both ampicillin and streptomycin must be used together to ensure that a copy of plasmid 1 is present in the bacterial cell and that the copy of plasmid 1 that is present contains an insert with the str^R gene of plasmid 2.

6. **e.** Digesting both plasmids with *Eco*RI yields a total of three fragments. Plasmid 1 is cut once, producing one linear fragment, whereas plasmid 2 is cut twice, giving rise to two linear fragments. One of the fragments from plasmid 2 will carry two antibiotic-resistance genes (tet^R and str^R), and the other will carry none. If the fragment that carries no antibiotic resistance genes ligates with the plasmid 1 fragment, the resulting plasmid would lose resistance to chloramphenicol due to the insertion of plasmid 2 DNA into the *Eco*RI site within the cam^R gene, but the plasmid would remain kan^R. This is the situation described in a. If the fragment from plasmid 2 carrying the tet^R and str^R genes is ligated to the plasmid 1 fragment, once again, resistance to chloramphenicol would be lost, but resistance to kanamycin would be maintained. However, in this case, the resulting plasmid would also gain resistance to tetracycline and streptomycin, the phenotype described in b. Ligation could also occur between the ends of each of the cut plasmids which would regenerate the original plasmids. Therefore, c is a possible phenotype because it could result from the recircularization of plasmid 1, whereas d is a possibility if the two fragments from plasmid 2 were to join back together. Only e is not a possibility, because for plasmid 1 to remain resistant to chloramphenicol, it cannot have a DNA fragment inserted into it, so it is not possible to get all four antibiotic resistance genes onto the same plasmid.

7. **b.** It is possible to lose resistance to ampicillin if a piece of mouse DNA is inserted into the *Bam*HI site, so it is possible to obtain a colony that is resistant to chloramphenicol and sensitive to ampicillin (cam^R amp^S) = a. Likewise, the vector can religate without taking in any mouse DNA, and the cell carrying this plasmid would be resistant to both antibiotics (cam^R amp^R) = c. If a bacterial cell does not take up a plasmid, it will remain sensitive to both of these antibiotics (cam^S amp^S) = d. This leaves b. If the cells are resistant to ampicillin, they would have to have taken up the plasmid with no insert and would be resistant to chloramphenicol also. It is therefore not possible in the current experiment to

recover cells resistant to ampicillin but sensitive to chloramphenicol (cam^S amp^R), which makes b incorrect.

TOPIC 3: ANALYZING AND USING CLONED GENES

KEY POINTS

- ✓ *How is the nucleotide sequence of a piece of DNA determined?*
- ✓ *How can a piece of one gene be used to detect related gene sequences in a different organism?*
- ✓ *What is meant by a restriction fragment length polymorphism?*
- ✓ *What is a DNA fingerprint?*

Once a bacterial clone carrying a particular DNA fragment has been identified, large amounts of that DNA fragment can be isolated by growing the cells in liquid culture in large fermentors. The cloned DNA can then be further characterized by restriction mapping or sequencing procedures. Both of these methods make use of a technique called **gel electrophoresis**.

GEL ELECTROPHORESIS SEPARATES DNA FRAGMENTS ACCORDING TO SIZE

In gel electrophoresis, DNA samples are loaded into wells near one end of a polymeric gel (e.g., agarose or acrylamide) that is supported on a tray immersed in buffer in an electrophoresis tank. Electrodes are attached at either end of the tank and a voltage applied across the gel. Because DNA carries a net negative charge, linear DNA restriction fragments migrate through the gel toward the positive electrode at a rate that is inversely proportional to their lengths. This is due to the fact that the polymeric fibers in the gel impede longer DNA fragments more than shorter fragments. Thus, over several hours, the DNA restriction fragments become spread out down the length of the gel according to size, forming a ladder of DNA bands that can be visualized by staining the DNA in the gel with a dye that fluoresces under ultraviolet light (e.g., the intercalating agent ethidium bromide; see Chapter 4). When an entire genome is used as the starting material, so many restriction fragments are produced that the bands appear as a smear on the gel but, for cloned pieces of DNA, the number of restriction fragments generated is typically of the order of 1–20, which are clearly visible as discrete bands.

RESTRICTION MAP SHOWS THE LOCATIONS OF RESTRICTION ENZYME RECOGNITION SITES

In **restriction mapping**, gel electrophoresis is used to separate by size the DNA fragments that result from treating a particular DNA molecule with a variety of restriction enzymes (first individually and then in all of the various combinations). The pattern of stained bands obtained on the gel is used to figure out where the restriction enzymes cut the DNA relative to one another. By studying the differences between the restriction maps of multiple alleles of a gene, scientists can often determine where a particular mutation has occurred within that gene or the extent of relatedness of different organisms.

NUCLEOTIDE SEQUENCE OF A PIECE OF DNA CAN BE DETERMINED

The maximum amount of information about a particular DNA fragment can be obtained by determining its complete nucleotide sequence. The most commonly used method for sequencing DNA is that of **dideoxyDNA sequencing**, which was developed by Frederick Sanger. This method relies on the mechanism of DNA replication. However, in addition to the normal components required for DNA replication *in vitro*, each dideoxy sequencing reaction also contains a small amount of a **dideoxyribonucleoside triphosphate (ddNTP)**. The dideoxy analogues are identical to the normal deoxyribonucleotides except that they lack the essential OH group at the 3′ carbon atom. As a result, a dideoxy analogue can be incorporated into a growing DNA chain, but the resulting chain cannot be further extended because there is no 3′ OH to which the next deoxyribonucleotide would normally be attached by a 5′ to 3′ phosphodiester bond. In this technique, multiple single-stranded copies of the piece of DNA to be sequenced are combined with DNA polymerase, a primer, and the four normal dNTPs (dATP, dCTP, dTTP, dGTP). The mix is then divided between four tubes, each containing a small amount of one of the four ddNTPs (each of which has been labeled with a different fluorescent tag). If each ddNTP is used at 1/100 of the concentration of the normal dNTP, an average of one dideoxy analogue will be incorporated into the newly synthesized DNA for every 100 of the normal deoxyribonucleotide added. In this way, each tube ends up containing a complex mixture of DNA fragments of differing lengths, where the length of each fragment identifies the position where a dideoxyribonucleotide was added to the growing chain (**Figure 7.7**). For example, consider the sequencing tube that contains ddTTP. Each time an A is encountered on the template strand, the DNA polymerase has the choice of adding either dTTP or ddTTP. If dTTP is added, synthesis continues. If ddTTP is added, chain growth terminates at that point and a truncated fragment is formed.

Because the tube contains multiple copies of the template DNA and both dTTP and ddTTP, a large number of truncated fragments will be formed, the lengths of these fragments corresponding to the locations of all of the As in the template strand. The contents of each sequencing tube are then denatured and separated according to size by gel electrophoresis. A laser beam that excites the fluorescent tags on the ddNTPs is used to mark the position of each fragment on the gel and by reading the gel one band at a time starting from the smallest, the entire sequence of the newly synthesized DNA can be determined. For example, in the figure below, the shortest fragment was made in the sequencing tube that contained ddATP. This means that the dideoxy analogue of dATP was responsible for the first truncation in the newly synthesized DNA and therefore that the first base must be an A. The second shortest fragment was made in the sequencing tube that contained ddGTP. Therefore, the second base in the newly synthesized

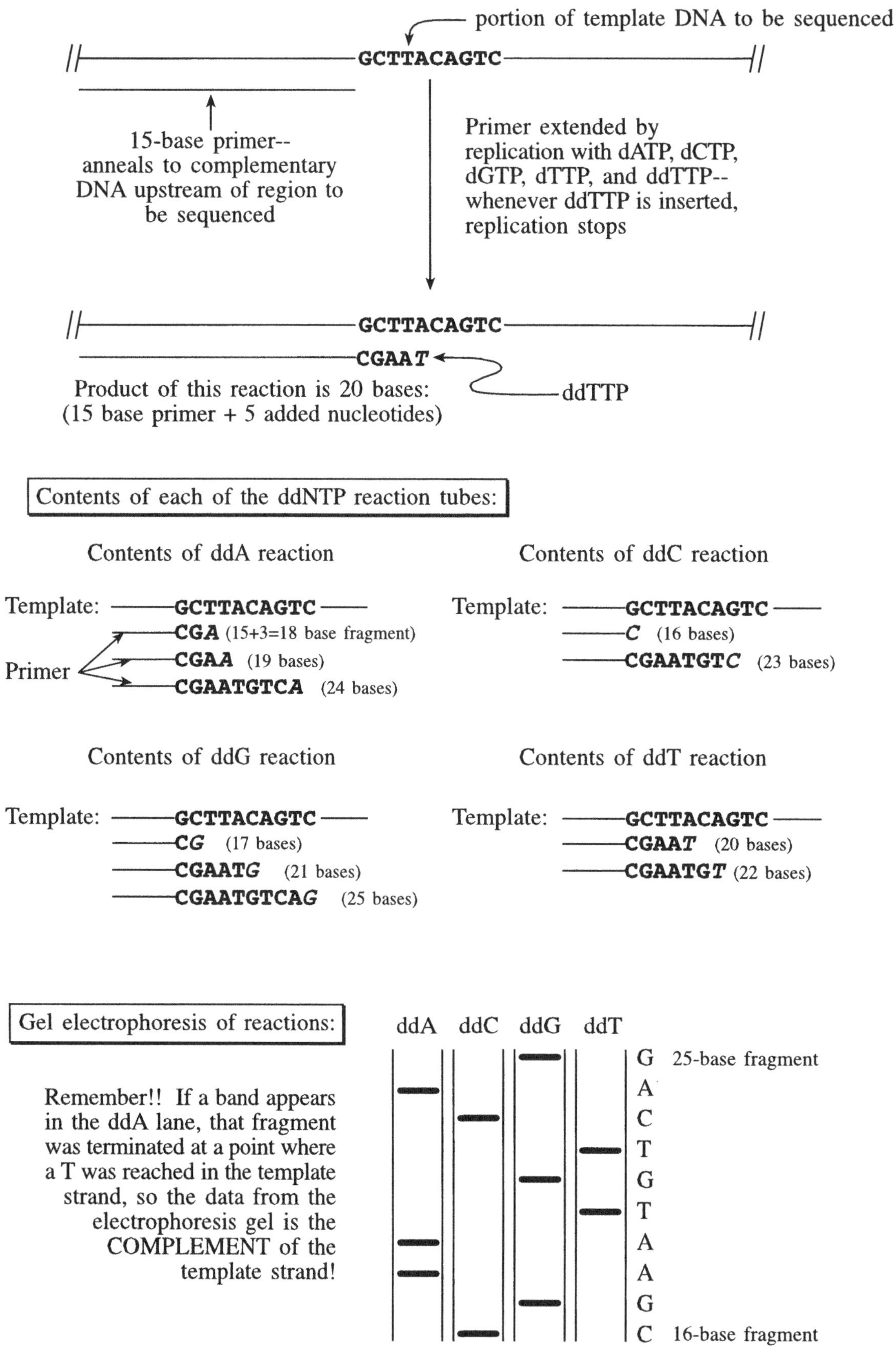

Figure 7.7 Dideoxy sequencing.

DNA must be a G. Using the same reasoning, the third base in the newly synthesized DNA is an A, the fourth a G, the fifth a C, and the sixth a T:

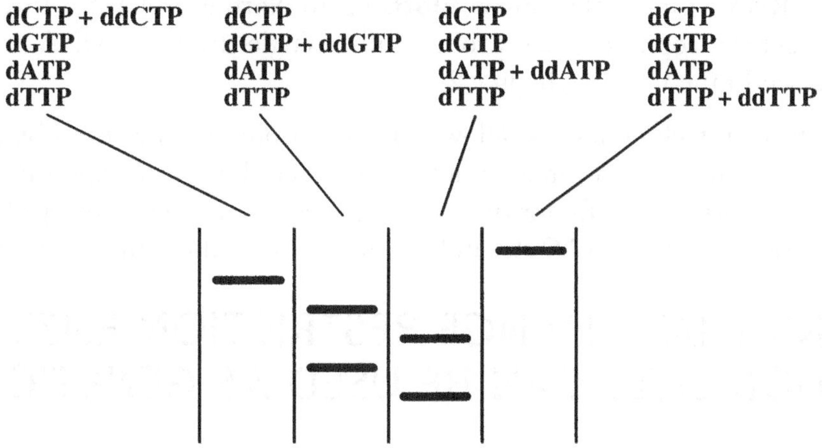

In this way, the entire sequence of the newly synthesized DNA can be read off the gel, and because this sequence is complementary to the template DNA, the sequence of the original DNA fragment can also be determined. This method has been used to determine the complete nucleotide sequence of more than 20 genomes (mostly bacterial) and by 2001, the human genome may have been completely sequenced using this technique.

DNA HYBRIDIZATION TECHNIQUES USE "PROBE" DNAS TO DETECT COMPLEMENTARY SEQUENCES

Many new genes arose during evolution by the duplication and divergence of existing genes and by the recombination of parts of different genes. Therefore, many genes have a family of close relatives (often of related function) elsewhere in the genome. To isolate each gene of a gene family from scratch using the cloning procedures described here would require a lot of effort, but this can be avoided once one member of a gene family has been isolated, by using a DNA hybridization technique called **Southern blotting**. In Southern blotting, sequences from the isolated gene are used as probes to search out complementary sequences among DNA fragments that have been transferred onto a solid support. Because of the sensitivity of DNA hybridization, the starting material for analysis can be entire genomic DNA. This DNA is cut with a restriction enzyme and the restriction fragments separated by gel electrophoresis.

After electrophoresis, the gel is placed on top of a sponge soaked in alkaline buffer, a sheet of nitrocellulose or nylon membrane laid on top of the gel, and paper towels stacked on top of the membrane. As the buffer is sucked up through the gel by the paper towels, the alkaline pH of the buffer denatures the double-stranded DNA fragments and the single-stranded DNAs are carried out of the gel and onto the nitrocellulose or nylon membrane where they adhere in bands exactly as they were in the gel. This process is referred to as "**blotting**" and the "blot" (the membrane carrying the single-stranded DNA fragments) is then peeled off the gel and sealed in a plastic bag with a buffer containing a radioactively labeled single-stranded DNA probe that is complementary to part of the desired DNA fragment.

The conditions under which the blot is exposed to the probe can be adjusted either to detect only those sequences that are perfectly complementary to the probe or to detect similar (homologous) sequences as well. To determine where the probe hybridizes to the blot, a sheet of photographic film is placed on top of the washed blot. The radioactivity in the bound probe exposes the film forming a dark band. An adaptation of this technique that is used to detect specific sequences in RNA is called **Northern blotting**. In Northern blotting, mRNA molecules are subjected to gel electrophoresis and probed with a single-stranded cDNA molecule or with antisense RNA (see In the Clinic, Chapter 2).

DNA hybridization techniques also allow sequences from one species to be isolated using probes derived from a related gene from a completely different but more experimentally accessible species. This is especially useful for determining evolutionary relationships between extinct and present-day species if enough DNA can be obtained from the remains of the extinct organism.

MUTATIONS THAT CHANGE RESTRICTION ENZYME RECOGNITION SITES CAN BE USED AS GENETIC MARKERS

Restriction enzymes cleave DNA at specific recognition sites. If a single base pair change occurs within a recognition site, the sequence can no longer be cut by the restriction enzyme. If a single base pair change that affects a restriction enzyme recognition site occurs in one of a pair of homologous chromosomes, fragments of different sizes are produced when the DNAs are digested with that particular enzyme. Differences between the DNA sequences of homologous chromosomes that result in different restriction fragment digest patterns are scattered throughout genomes, including the human genome. Mutated restriction enzyme sites that occur frequently in a population and produce distinctive patterns of DNA fragments are called **restriction fragment length polymorphisms**, or **RFLPs** (pronounced "riflips"), and they are often used as genetic markers for a particular location in the genome. RFLPs can be readily observed by gel electrophoresis of restricted DNA and Southern blotting with an appropriate marker. A simple example of an RFLP is illustrated by the sickle cell anemia mutation. For this recessive disease, the exact nucleotide change in the mutant gene is known; the sequence GAG that codes for the amino acid glutamic acid is changed to GTG that codes for valine. This nucleotide change also eliminates one of the recognition sites for the restriction enzyme *Mst*II. The incredible specificity of DNA hybridization techniques makes it possible to detect this mutation. Thus, when DNAs from sickle cell and "normal" individuals are cut with *Mst*II, the fragments separated by gel electrophoresis, and the DNAs hybridized with a probe that recognizes the *Mst*II fragment affected by the mutation, the probe DNA will bind to larger sized fragments in the presence of the mutation, allowing individuals that carry one or two copies of the sickle cell mutation to be distinguished from "normal" individuals.

PRACTICAL APPLICATIONS OF DNA TECHNOLOGY

Differences in Restriction Enzyme Profiles Can Be Used to Identify Individuals

DNA fingerprinting or **DNA typing** is used to characterize biological evidence left at crime scenes to establish whether a suspect could have committed a crime. In this technique, DNA is isolated from the biological evidence (e.g., blood, semen, skin, hair), cut with a restriction enzyme, and the fragments separated by gel electrophoresis and transferred by Southern blotting onto a nylon membrane. The blot is then hybridized consecutively with different radiolabeled probes that each recognize a distinct DNA sequence and the hybridization pattern obtained with each probe visualized using x-ray film which is exposed by the radioactivity. The most commonly used probes for this type of analysis consist of human minisatellite DNAs. Minisatellite DNAs are short tandemly repeated sequences that occur throughout the human genome. The lengths of the repeats range from 9 to 40 base pairs and the number of repeats usually ranges from 10 to 30. DNA fingerprinting is based on the observation that the same minisatellite DNA region typically contains a different number of repeats in different individuals. A minisatellite DNA fingerprint represents the lengths of a variety of these repeat regions. Because of the extensive variability in human minisatellite DNA regions, the chances of finding two individuals with the same DNA fingerprint is about one in 10^5 to 10^8. In other words, a DNA fingerprint based on minisatellite DNA sequences is about as unique as an individual's fingerprints.

Recombinant DNA Techniques are Transforming the Medical Industry

Recombinant DNA techniques are making widespread and important contributions to medicine. For example, DNA hybridization studies, using labeled probes to detect RFLPs, have opened a new chapter in disease diagnosis. Genes have already been isolated for many common human diseases, including cystic fibrosis, Huntington's, hemophilia, and sickle cell anemia, and by using relevant parts of these genes as probes, individuals with these genetic diseases can now be identified before the onset of symptoms, as discussed previously. Even in cases where the gene associated with a particular disease has not been cloned, the presence of the disease gene can be determined with a reasonable degree of accuracy if the disease gene is closely linked to (and hence presumably coinherited with) an RFLP.

Many medically important proteins are now produced using recombinant DNA techniques. By transferring human genes into bacterial or yeast cells, using what are now routine cloning procedures, it is possible to make large quantities of human proteins that are present naturally in only minute amounts. Proteins made this way include human insulin, human growth hormone, blood clotting factors, tissue plasminogen activator, numerous recombinant vaccines, and anticancer drugs, to name a few.

For any genetic disease caused by a single defective allele, it should theoretically be possible, using recombinant DNA techniques, to replace the defective allele with a normal copy of that gene. This is the premise behind **gene therapy**, one of the most controversial applications of recombinant DNA technologies. Gene therapy requires that the defect or defects responsible for a specific disease be known and that techniques be developed for the efficient transfer of DNA into human cells. Considering the rapid rate of progress on the Human Genome Project (now projected to be finished by 2001), it is highly likely that there will be a dramatic increase in the identification of specific defects in known and cloned genes. The advances that are being made in the identification of defects associated with specific diseases, coupled with improvements in gene delivery vectors (see In the Clinic, Chapter 1), increase the probability that gene therapy will eventually lead to cures for a number of devastating human diseases.

Recombinant DNA Techniques are Revolutionizing Agricultural Practices

The creation of new varieties of plants by the direct manipulation of their genotypes is an increasingly important application of recombinant DNA technologies. For example, most of the soybean crops grown in the United States are "Roundup-ready," that is, they are resistant to the herbicide glyphosate (known by the trade name Roundup). Glyphosate kills plants by inhibiting 5-enolpyruvylshikimate 3-phosphate synthase (EPSPS), a chloroplast enzyme that is involved in the synthesis of essential amino acids. Fortunately, some soil bacteria contain a mutant form of this enzyme that is much less sensitive to glyphosate. Recombinant DNA technologies were used to isolate the mutant EPSPS gene and to modify it so that it could be expressed in plants. Transgenic Roundup-resistant (or "ready") cotton and soybean plants were then produced by *Agrobacterium*-mediated gene transfer. *Agrobacterium tumefaciens* is a soil bacterium that causes a plant disease called crown gall, characterized by the formation of tumors in the plant tissues. *Agrobacterium* is a natural genetic engineer because it transfers part of its own DNA (the so-called T-DNA) into the genome of its host plant. The T-DNA resides on the Ti (for tumor inducing) plasmid and it was shown that any DNA inserted into the T-DNA region can be transferred into the host cell's genome. Therefore, the Ti plasmid of *Agrobacterium* has been exploited as a vector to transfer foreign genes into plants.

Agrobacterium-mediated gene transfer has also been used to produce plants that make their own insecticides. Some bacteria produce proteins that kill the insect larvae that eat them, and in the case of some strains of *Bacillus thuringiensis*, the toxicity of these insecticidal proteins is 80,000 times that of commonly used commercial insecticides. The toxin genes from different strains of *Bacillus thuringiensis* were isolated, fused to the necessary plant regulatory sequences, and introduced into plants using the Ti plasmid vector of *Agrobacterium tumefaciens*. In this way, transgenic potato, cotton, tomato, and corn crops with considerable resistance to their insect predators have been produced.

In addition to the generation of herbicide-, insecticide-, and insect-resistant plants, much effort in plant genetic engineering has been directed toward the improvement of properties such as nutritional value and postharvest quality (see In the Clinic, Chapter 2). Recombinant DNA technologies have also been used to produce plants capable of synthesizing useful products such as antibodies and biodegradable plastics.

In animal husbandry, recombinant DNA technologies are being used to produce farm animals that grow faster or produce more milk or leaner meat or that can make medically important products (e.g., transgenic goats that produce tissue plasminogen activator, a protein that dissolves blood clots, in their milk). In 1997, Ian Wilmut and his colleagues sparked a worldwide debate over the moral and medical implications of cloning by producing the first cloned animal. Nuclei from fibroblast cells taken from the udder of an adult sheep were transplanted into enucleated eggs (i.e., eggs in which the nucleus had been destroyed) from a different sheep and the eggs implanted into the uterus of a surrogate mother. The result, a lamb named Dolly, was genetically identical to the sheep from whose udder the donor nucleus had been obtained. Wilmut's experiment could revolutionize the production of transgenic animals. For example, by cloning, many identical "copies" of sheep that have been genetically engineered to produce a valuable pharmaceutical in their milk could be made.

Environmental Uses of Recombinant DNA Technologies

Increasingly, genetically engineered microorganisms are being used to solve environmental problems. For example, many bacteria can extract minerals such as copper and lead from the environment and incorporate them into compounds that can be easily recovered. Such bacteria are being genetically modified so that they can be used to mine minerals or clean up toxic mining wastes. Yet other genetically modified microorganisms are being used to degrade some of the compounds released by sewage waste and oil spills. As biotechnologists identify more microorganisms capable of detoxifying specific waste products, more strains will be developed that will be able to survive the harsh conditions of the environmental problem (be it a slag heap from an old mine or an oil spill) and still be able to detoxify the waste products.

Topic Test 3: Analyzing and Using Cloned Genes

True/False

1. In the technique known as Southern blotting, DNA restriction fragments are separated by gel electrophoresis, transferred to a special membrane, and hybridized to a radioactively labeled DNA probe.

2. Knowing the DNA sequence of a bacterial gene would give you precise information about the amino acid sequence.

3. Knowing the amino acid sequence of a bacterial protein would give you precise information about the DNA sequence.

Short Answer

4. Multiple restriction enzymes are used to generate a collection of overlapping DNA fragments which can be used to generate a ____________.

5. Define RFLP.

Multiple Choice

6. Which of the following statements about molecular techniques is *incorrect*?
   a. In the technique of gel electrophoresis, large DNA fragments end up closer to the bottom of the gel (farthest from the wells) because they carry a greater negative charge than smaller DNA fragments.
   b. The dideoxy approach to DNA sequencing uses dideoxy analogues of the normal deoxyribonucleoside triphosphates (dNTPs) to terminate DNA synthesis at all positions along the DNA template.
   c. PCR technique is used to amplify specific DNA sequences.
   d. Differences in the lengths of fragments obtained by digesting the DNAs of different individuals with a variety of different restriction enzymes explains the basis behind using restriction enzymes to assign a unique DNA (genetic) fingerprint to every individual.
   e. In cloning, a plasmid vector is used to make many identical copies of a DNA fragment.

7. If you were told that the 3′ to 5′ sequence of a single-stranded piece of DNA was 3′-CGACCTATTG-5′, which of the following accurately depicts the dideoxy sequencing gel that would be obtained from sequencing this piece of DNA?

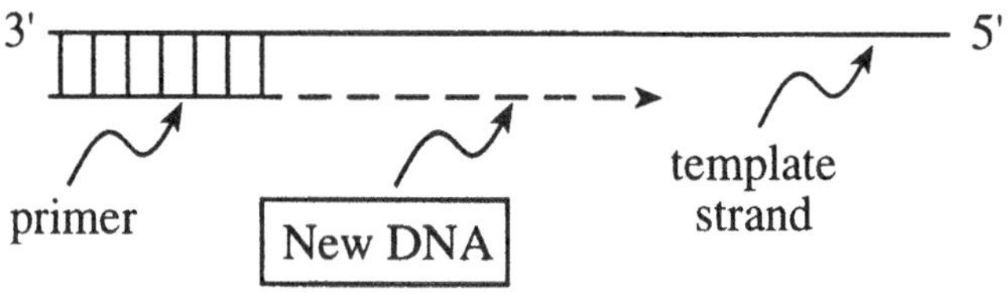

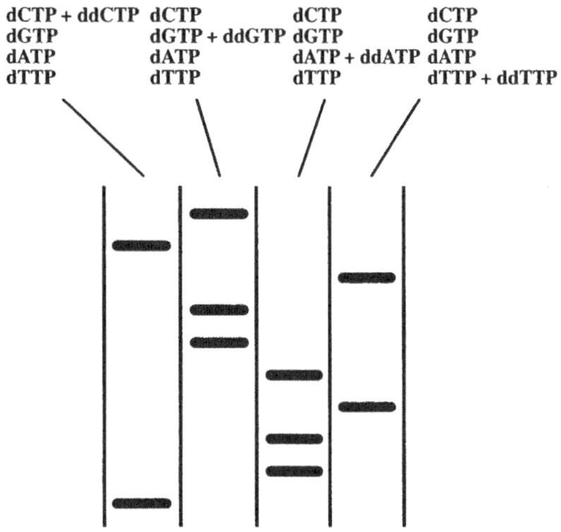

a.

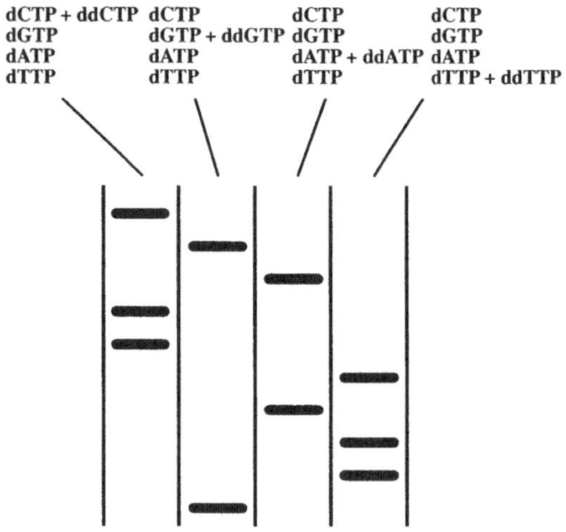

b.

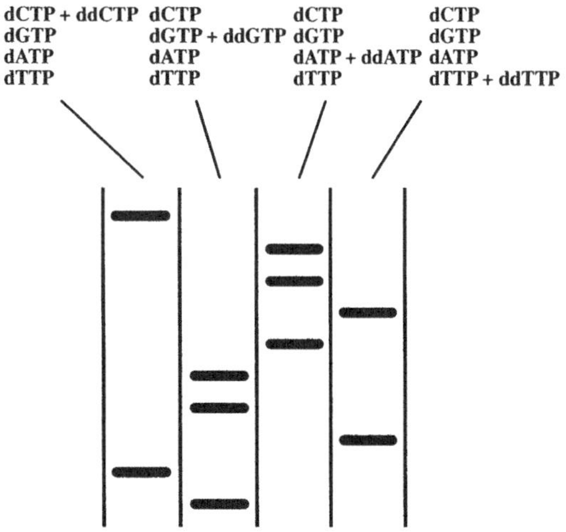

c.

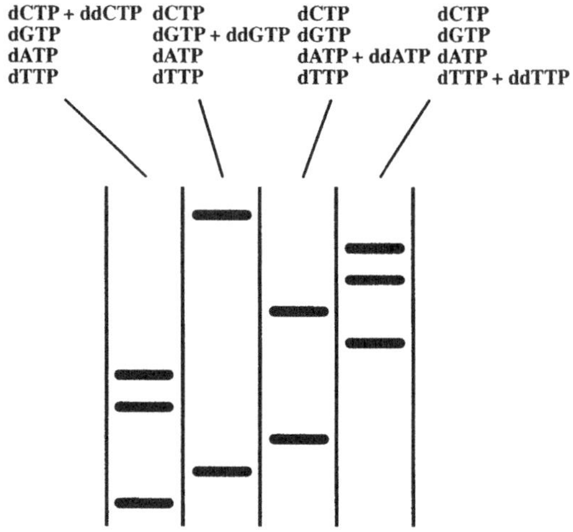

d.

e. None of the above accurately depicts the dideoxy sequencing gel that would be obtained from sequencing this piece of DNA.

8. The nucleotide sequence of a DNA fragment was determined by the dideoxy DNA sequencing method. The data are shown below:

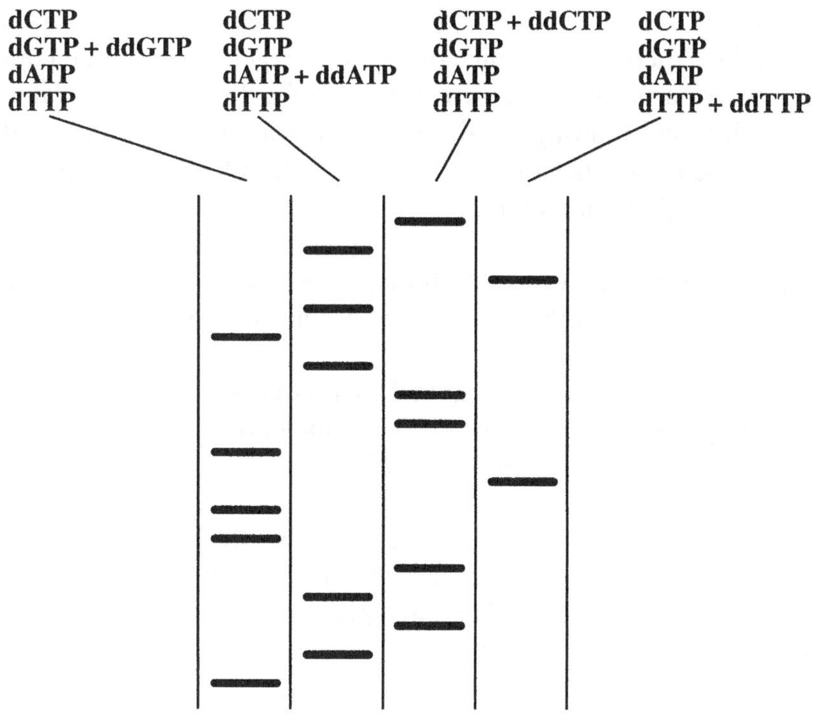

What is the **5′ to 3′** sequence of nucleotides in the **template** DNA strand (the template DNA is the single-stranded DNA that is present in each of the sequencing tubes at the start of the reaction—the template DNA is *not* the DNA that is synthesized).

a. GACACGGTGCCAGATAC
b. CATAGACCGTGGCACAG
c. CTGTGCCACGGTCTATG
d. GTATCTGGCACCGTGTC
e. GACACGGUGCCAGAUAC

Topic Test 3: Answers

1. **True.** In Southern blotting, a short piece of DNA is used to "probe" for complementary gene sequences in the same or in a different organism's genome. The single-stranded probe DNA can base pair to complementary single-stranded sequences that have been immobilized on the membrane and, by labeling the probe in some way, the location of bound probe can be determined.
2. **True.** Bacterial genes typically do not contain introns, so all of the information in the gene is directly transcribed into mRNA. Because the genetic code is known, it is possible, once the initiator codon has been located, to determine what amino acid will be added for each codon downstream of the initiator codon, giving the exact amino acid sequence of the protein. The initiator codon is located by finding the first methionine codon (AUG) after the Shine-Dalgarno sequence, and where translation ends is determined by finding the first stop codon in the correct reading frame.

3. **False.** The degeneracy in the genetic code means that it is impossible to figure out the precise DNA sequence of the corresponding gene even if the complete amino acid sequence of a protein is known. For example, if the second amino acid is a tyrosine (the first amino acid is always methionine), all that could be concluded is that the codon in the mRNA could have been either UAU or UAC. Other amino acids can be specified by as many as six different codons.

4. **Restriction map.** A restriction map shows where restriction enzymes cut a DNA molecule relative to other restriction sites in the DNA.

5. A difference between homologous DNA sequences that leads to a variation in the lengths of the fragments produced when the homologous DNAs are digested with the same restriction enzyme is called an RFLP.

6. **a.** All DNA is negatively charged and will be drawn toward the positive electrode, but the gel matrix (e.g., agarose or acrylamide) hinders the movement of the DNA fragments through the gel. Smaller DNA fragments can pass through the gel matrix more easily than larger DNA fragments and therefore will move further down the gel in a given amount of time than larger fragments.

7. **c.** The sequence of the template DNA is 3′-CGACCTATTG-5′. Because DNA synthesis only occurs in the 5′–3′ direction and the double helix is antiparallel, the first base added to the complementary strand would be a G, followed by C, T, G, G, as follows:

   3′-CGACCTATTG-5′
   5′-GCTGGATAAC-3′

   Because the sequencing gel records the order of ddNTP incorporation (and therefore chain termination), the shortest possible fragment (in other words, the earliest termination) would be expected to occur in the lane containing the contents of the tube with ddGTP. This fragment, being the shortest, would also travel the farthest, so the band at the bottom of the gel would be in the ddGTP lane. This eliminates a and d, which suggests that the first base incorporated would be a C. Continuing down the sequence, the next base incorporated would be a C, followed by a T. This eliminates b, which suggests that the second base incorporated would be a T followed by another T. Only c gives the correct complement to the template DNA sequence (5′-GCTGGATAAC-3′).

8. **d.** The 5′ to 3′ sequence of the DNA synthesized among the 4 sequencing reactions reading the gel from bottom to top is 5′-GACACGGTGCCAGATAC-3′. Because the newly synthesized DNA is complementary and antiparallel to the template DNA as follows:

   3′-CTGTGCCACGGTCTATG-5′ = template DNA
   5′-GACACGGTGCCAGATAC-3′ = newly synthesized DNA

   the **5′ to 3′** sequence of the template DNA is 5′-GTATCTGGCACCGTGTC-3′ (d).

IN THE CLINIC

Once the cause of a specific genetic defect has been identified, it is possible to detect the presence of that defect in an individual's genotype, often well before any symptoms appear (e.g., many women in families with a history of breast cancer are now routinely screened for the *BRCA1* and *BRCA2* "breast cancer" genes). With the completion of the Human Genome Project, genetic screening for many more (4,000+) genetic defects will be possible. This future possibility has sparked a worldwide debate over who should be allowed access to an individual's genetic information. Thus, should an individual's employer have access to his or her genetic information? What about the individual's insurance company (would an HMO want to insure a person if a genetic test indicated that person had inherited a lethal disease such as Huntington's)? Will individuals with "flawed" genetic profiles be discriminated against? This is the moral behind the movie "Gattaca"; in this movie people are segregated on the basis of their genetic profiles. Obviously, "Gattaca" is an extreme (and hopefully incorrect) view of the possible implications of genetic testing, but in real life the potential for abusing the information exposed by genetic tests exists, and this is the reason why approximately 4% of the investment in the Human Genome Project—about $90 million!—is going to studies seeking to untangle the ethical and social issues raised by this new era of genetic testing.

The good news is that the time is coming when simple tests on a minuscule sample of blood or tissue will be able to reveal the presence of potentially harmful diseases in time to eradicate many of them or to take preventative measures to slow their onset. For example, leading a healthy lifestyle, by eating a balanced, high-fiber diet rich in fruit and vegetables, and following a daily exercise plan, can delay or prevent heart disease, even in those individuals who are more susceptible to it due to a genetic defect. That raises the question of whether genetic screening should be limited to those diseases or disorders for which some treatments are currently available. What about inherited diseases? Should individuals be screened for these disorders so that they can make informed decisions about their future choices, including whether or not to have children? Should "high-risk" individuals be required to have a genetic test? All these questions and more are being asked as the ability to make predictions about an individual's future health based on his or her genetic makeup becomes reality.

DEMONSTRATION PROBLEM

You want to use *E. coli* to produce large amounts of a functional human protein that appears to have anticancer properties. You know that the protein is produced by the pituitary. How would you go about cloning the gene for this protein so that it can be expressed at high levels in *E. coli*?

Using the recombinant DNA techniques discussed in this chapter, it is possible to produce large quantities of many human proteins in *E. coli*. Ideally, we would like to simply isolate the anti-cancer gene from human genomic DNA and transfer it into an *E. coli* expression system.

However, the first problem we would encounter is the likely presence of introns in the human DNA, as illustrated in the figure below:

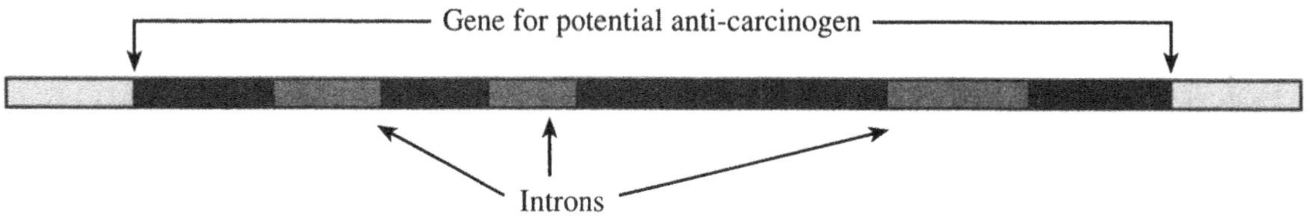

E. coli does not have the ability to recognize and remove introns, so if this gene was directly isolated from human genomic DNA and expressed in *E. coli*, a nonfunctional protein would be produced because noncoding sequences and coding sequences would be transcribed and translated. Therefore, because the goal of this cloning is to produce a functional human protein in bacteria, starting with genomic DNA is not an option. As discussed in this chapter, the other source of genes for cloning is complementary DNA, or cDNA. A cDNA **library** is constructed by isolating all of the mRNAs present in a given cell type and using these to prepare DNA copies of the genes without the introns.

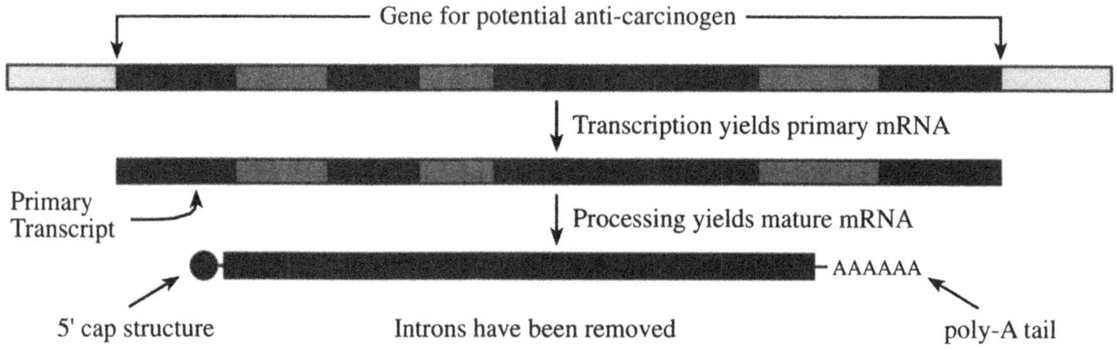

Because the expression of most eukaryotic genes is cell specific, mRNA needs to be isolated from cells in which the protein of interest is being produced. Fortunately, we know that our potential anticarcinogen is produced by the pituitary, so total cell mRNA would need to be isolated from this tissue. Mature mRNAs are isolated from other cell components (including genomic DNA, transfer RNAs, ribosomal RNAs, and primary mRNA transcripts) by exploiting their poly(-A) tails. Thus, total pituitary cell extracts are poured over a column filled with a poly(-T) matrix to which poly(-A) tails bind tightly. The column is then washed with a series of solutions to ensure that all of the other cell components are removed, followed by a final solution to release the mature pituitary mRNAs from the matrix. Using this approach, the mRNAs produced in pituitary tissue, including the mRNA for the anticancer gene, could be isolated.

The next step in constructing a cDNA library is to prepare cDNAs from the purified mRNAs. Once again, the poly(-A) tail of mRNA molecules is exploited to prepare a primer that is needed to prime synthesis of DNAs from the mRNA templates. DNA synthesis using RNA as a template requires a special enzyme, called **reverse transcriptase**, that is isolated from retroviruses. DNA synthesis by reverse transcriptase yields a DNA–RNA hybrid, which would be of no use in an expression system. The DNA–RNA hybrid is therefore treated to degrade the RNA component, leaving a single-stranded DNA molecule, which is then replicated by DNA polymerase to form a complete double-stranded cDNA molecule.

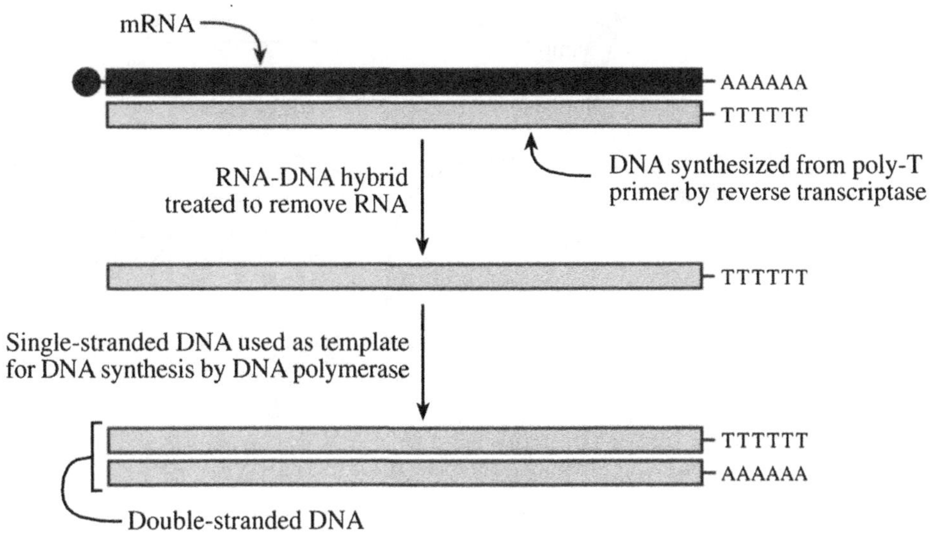

The resulting double-stranded cDNAs must now be inserted into an expression vector. To do this, small fragments of DNA called **linkers**, which contain known restriction enzyme recognition sequences, are ligated to the ends of the cDNAs as shown in the figure below (obviously you should choose a restriction enzyme recognition site that is not found in your target gene and that cuts only once in your expression vector):

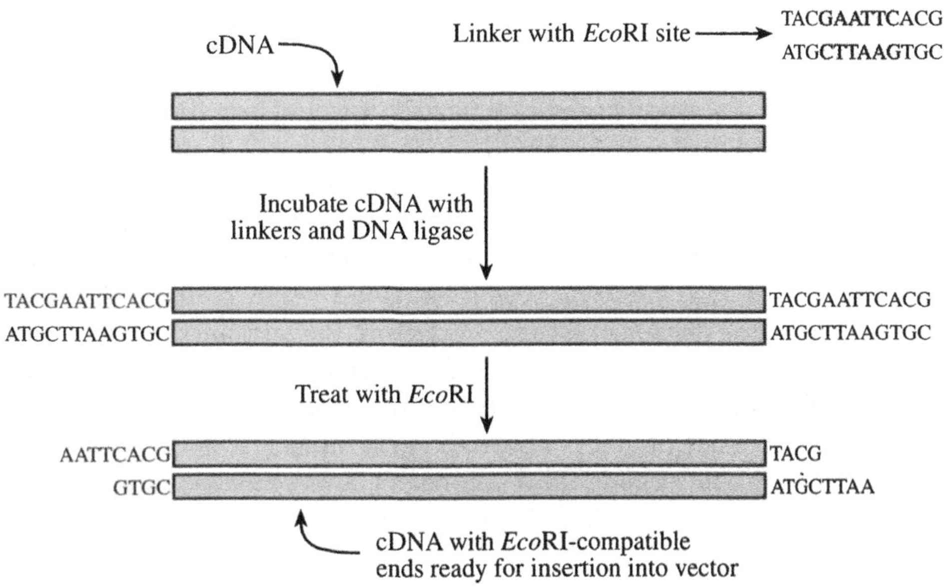

The choice of expression vector is also important. Because cDNAs have neither promoters nor transcription terminators, a vector containing both of these regulatory sequences needs to be used. Many of these so-called **expression vectors**, with unique restriction sites between strong promoter and terminator sequences, have been constructed. Inserting the pituitary cDNAs into such a vector, as illustrated below, would allow their expression in *E. coli*:

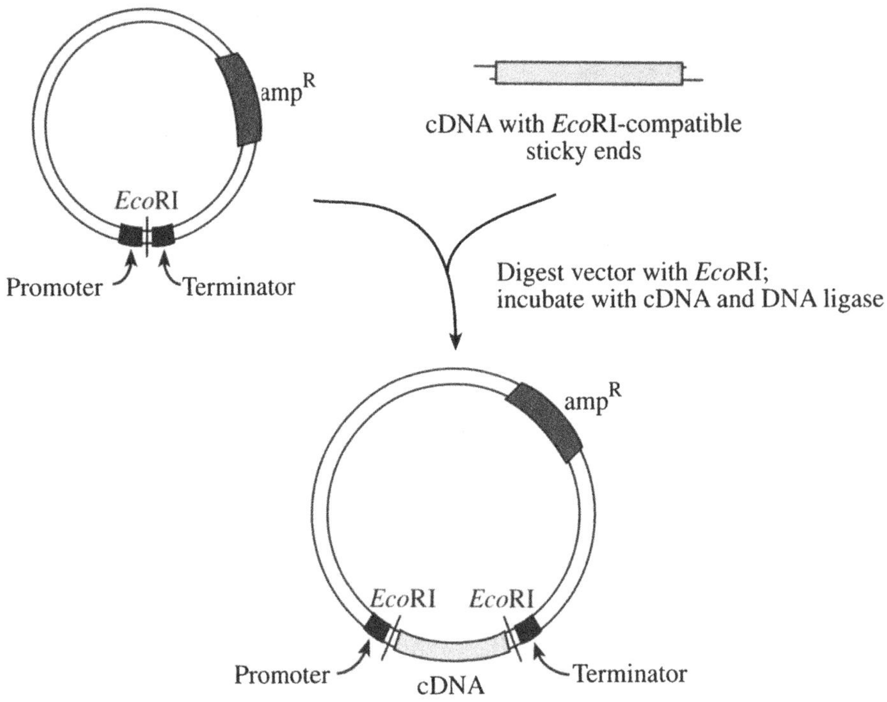

Transforming these resulting plasmids into an *E. coli* strain and selecting for ampicillin resistance (amp^R) would generate a cDNA library, with plasmids representing all the protein-coding genes expressed in pituitary tissue. Additional experiments would need to be performed to identify plasmids carrying the cDNA that codes for the desired protein. This could be done by using a piece of the anticancer gene as a "probe" to search out complementary sequences in the DNAs prepared from the cDNA library (see the discussion on DNA hybridization techniques in Topic 2 of this chapter).

Unless expression of the cloned gene is regulated, it would be expressed continuously in the *E. coli* host strain, and at elevated levels due to its presence on a multicopy plasmid with a strong promoter. However, it is possible that the anticancer protein could be toxic to the *E. coli* cell. If this turned out to be the case, expression of the human gene could be regulated. For example, as discussed in the bacterial regulation chapter, expression of the lactose operon is tightly regulated. If the expression vector contained a copy of the *lacI* gene and the promoter and operator sequences for the *lac* operon, the anticancer gene would not be expressed at high levels in the absence of inducer [lactose, or a gratuitous inducer like isopropyl-thiogalactoside (IPTG), which mimics the action of lactose as an inducer but is not a substrate for β-galactosidase]. The addition of inducer would promote high levels of expression of the anticancer gene. This approach allows potentially toxic genes to be maintained in the host strain without killing it. Alternatively, a eukaryotic expression system, such as yeast, could be used.

Cloned gene proteins are much easier to purify if they are secreted or exported out of the host cell. Therefore, to make purification of the anticancer protein easier, it could be fused to special signal or leader peptide sequences that would target the protein for export from the cell. Specialized vectors carrying unique restriction enzyme recognition sites immediately adjacent to leader peptide sequences are available.

Chapter Test

True/False

1. Dideoxyribonucleoside triphosphates (ddNTPs) terminate DNA synthesis because they lack a 2′ hydroxyl (–OH) group on the di deoxyribose residue.
2. For the polymerase chain reaction, it is necessary to know some of the sequence flanking the 3′ ends of the double-stranded DNA sequence to be amplified.
3. A restriction map shows where different restriction enzymes cut a DNA molecule relative to other restriction sites in the DNA.

Short Answer

4. Match each of the recombinant DNA "tools" listed on the left below with one of the applications listed on the right. Each tool/application should be used once only.

1 = Reverse transcriptase	a = Generation of recombinant DNAs
2 = Restriction enzymes	b = Determination of the base sequence of DNA fragments
3 = Dideoxyribonucleoside triphosphates	c = Amplification of specific DNA sequences
4 = Heat-stable DNA polymerase	d = Production of eukaryotic genes that can be expressed in bacteria
5 = DNA ligase	e = Fragmentation of large DNA molecules

5. The process of using a plasmid vector to make many identical copies of a piece of DNA in a host cell is referred to as ____________.
6. ____________ obtained from biological evidence found at the crime scene can be used to eliminate suspects from criminal investigations.

Multiple Choice

7. You are asked to prepare a cDNA library containing the gene for human insulin, which is expressed in the pancreas. Which of the following would you ***not*** need to isolate in order to prepare this library?
   a. Pancreatic cell DNA
   b. Pancreatic cell mRNAs
   c. DNA polymerase
   d. Reverse transcriptase
   e. DNA ligase
8. Which of the following do the PCR, dideoxy DNA sequencing, Southern blotting, and sticky end ligations have in common?
   a. They all involve the synthesis of new DNA.
   b. They all involve complementary base pairing.
   c. They all involve denaturation of DNA.
   d. They all result in the amplification of a piece of DNA.
   e. They all involve DNA ligase.
9. Restriction enzyme 1 cuts at all sites cut by restriction enzyme 2, but restriction enzyme 2 cannot cut at any of the sites for restriction enzyme 1. This is possible if

a. restriction enzyme 1 can recognize and cut at more than one site (i.e., it is a nonspecific restriction enzyme).
b. restriction enzyme 1 recognizes a six base-pair sequence, whereas restriction enzyme 2 recognizes a four base-pair sequence which is identical to the internal four base-pair sequence of the site recognized by restriction enzyme 1.
c. restriction enzyme 2 recognizes a six base-pair sequence, whereas restriction enzyme 1 recognizes a four base-pair sequence that is identical to the central four base-pair sequence of the site recognized by restriction enzyme 2.
d. restriction enzymes 1 and 2 recognize the same sequence of bases but the bases in the site for restriction enzyme 1 are methylated which prevents cutting by restriction enzyme 2.
e. None of the above

10. Which one of the following statements about plasmids is ***incorrect***?
   a. The fact that plasmids contain their own origin of replication means that they are able to replicate outside of a host cell.
   b. The presence of a plasmid in a bacterium is frequently determined by testing for resistance to a particular antibiotic.
   c. Plasmids can be introduced into bacterial host cells by a process known as transformation.
   d. Plasmids often contain multiple cloning sites (or polylinkers) that consist of a small region of DNA containing a cluster of unique recognition sequences for a variety of restriction enzymes.
   e. Many of the commercially available plasmids have been specially engineered so that they can make hundreds of copies of themselves.

11. Which of the following statements is ***incorrect***?
   a. Blunt-ended fragments obtained by cutting two different DNAs can only be joined together if the same restriction enzyme was used to cut both DNAs.
   b. Usually, DNA fragments with sticky ends can only be joined together without any further treatment if the same restriction enzyme was used to generate these fragments.
   c. Restriction enzymes are produced naturally by some bacteria to degrade invading foreign DNA.
   d. Restriction enzymes can only recognize and degrade double-stranded DNA if their recognition sequences are unmethylated.
   e. Most sites recognized by restriction enzymes read the same from left to right across the top DNA strand as from right to left across the bottom DNA strand (i.e., they are palindromes).

Chapter Test Answers

1. **False**
2. **True**
3. **True**
4. **1 = d, 2 = e, 3 = b, 4 = c, 5 = a**
5. **Cloning**

6. **DNA fingerprints**

7. **a** 8. **b** 9. **c** 10. **a** 11. **a**

Check Your Performance

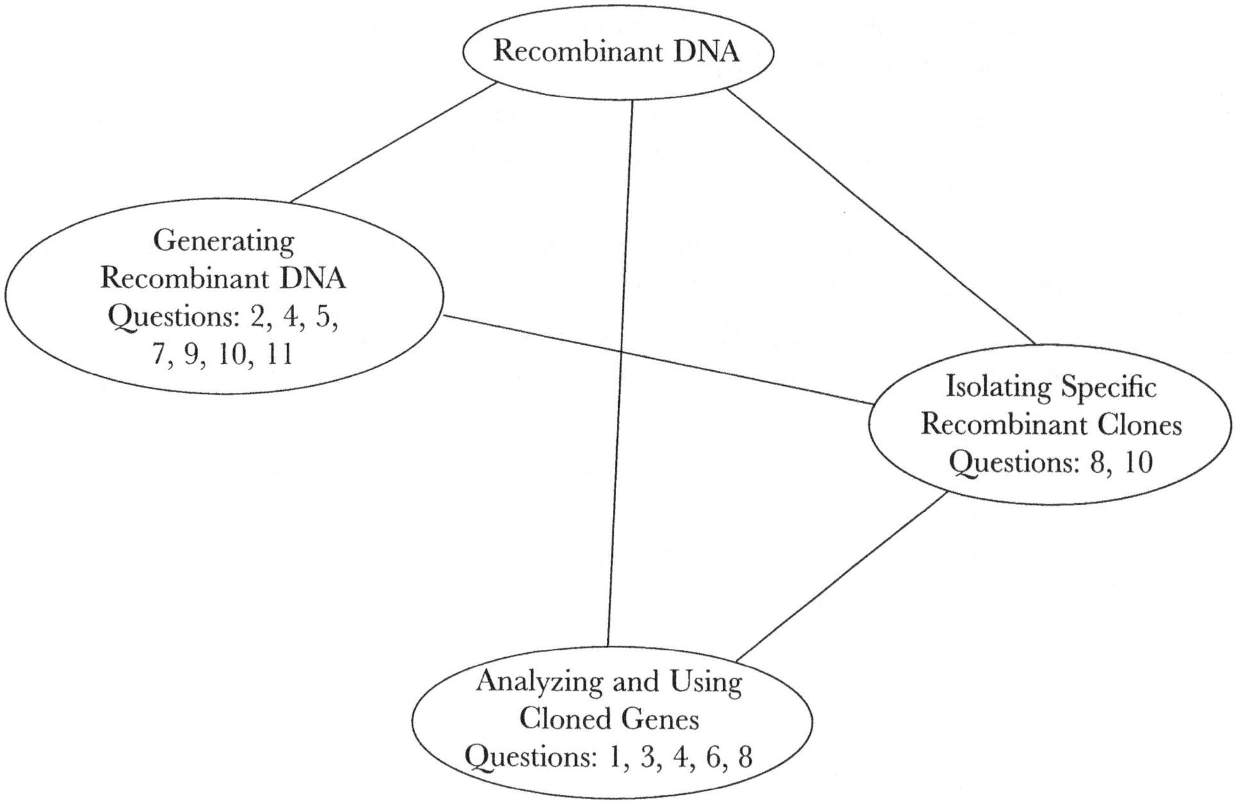

Note the number of questions in each grouping that you got wrong on the chapter test. Identify where you need further review and go back to relevant parts of this chapter.

Final Exam

True/False

1. In the Ames test, rat liver extracts are used to mimic the metabolic changes a potential mutagen would experience in the human body.
2. The primary role of gene regulation in bacteria is to ensure that the right gene is expressed in the right cell type.
3. In eukaryotes methylation of nucleotides in DNA plays an important role in development by keeping genes that have already been turned off from being turned back on.
4. Eukaryotic cells have evolved alternative modes of RNA splicing as a strategy to produce the same protein from many different genes.
5. A complementary DNA (cDNA) library is a collection of recombinant plasmids containing DNA inserts corresponding to all of the mRNAs purified from a particular tissue.
6. The technique of dideoxy DNA sequencing requires a heat-stable DNA polymerase.
7. The band pattern obtained on a dideoxy sequencing gel directly represents the nucleotide sequence of the DNA strand that is synthesized as a series of truncated fragments in the sequencing tubes.

Multiple Choice

8. Which one of the following statements about the DNA double helix is correct?
   a. In a DNA double helix, the number of guanines divided by the number of adenines = 1.
   b. If a DNA double helix is 30% cytosine, it is 40% thymine.
   c. The three-dimensional structure of a DNA double helix is stabilized by hydrogen bonds between complementary base pairs and base-stacking interactions.
   d. The DNA double helix consists of two DNA chains that wind around one another in such a way that a purine in one of the chains is always opposite a purine in the other chain and vice versa.
   e. Watson and Crick used Chargaff's x-ray data to determine the three-dimensional structure of DNA.
9. The conversion of a double-stranded DNA molecule to single strands (denaturation) is influenced by
   a. the temperature of the DNA solution.
   b. the pH of the DNA solution.
   c. the concentration of positively charged ions such as Na^+ in the DNA solution.
   d. the percentage of GC base pairs in the DNA molecule.
   e. All of the above.
10. Which one of the following statements about DNA replication is correct?
   a. When a double helix replicates, Okazaki fragments are formed on one of the two parental DNA strands only.

b. A single RNA primer is needed to initiate each leading strand whereas multiple RNA primers are needed for the synthesis of a complete lagging strand.
c. DNA replication is referred to as "semiconservative" because both of the parental strands end up in the same daughter cell.
d. DNA replication is initiated at multiple sites in bacterial genomes but at only one site in eukaryotic genomes.
e. DNA synthesis occurs in the 5′ to 3′ direction on one side of each *ori* and in a 3′ to 5′ direction on the other side of each *ori*.

11. In the Meselson-Stahl experiment, *E. coli* cells grown in the presence of a "heavy" isotope of nitrogen (^{15}N) were transferred into fresh medium containing normal "light" nitrogen (^{14}N). The cells were allowed to go through 3 cell divisions in the "light" medium and DNA was isolated after each cell division and run on cesium chloride density gradients. What would the cesium chloride density gradients have looked like after three cell divisions in the light (^{14}N) medium? [Note: the "thickness" of a band is indicative of the number of DNA molecules banding at that position—the thicker the band, the greater the number of DNA molecules.]
a. One band in the middle of the gradient
b. One band at the top of the gradient
c. Two bands of equal thickness; one in the middle of the gradient and one at the top of the gradient
d. Two bands; a thinner one in the middle of the gradient and a thicker one at the top of the gradient
e. Two bands; a thicker one in the middle of the gradient and a thinner one at the top of the gradient

12. If the 5′ to 3′ sequence of one strand of a DNA double helix is

5′-TCCGATGC-3′

the 5′ to 3′ sequence of the opposite DNA strand would = X

If X is the DNA sequence that is used as a template by RNA polymerase, the 5′ to 3′ sequence of the resulting mRNA would = Y, where:
a. X = 5′-AGGCTACG-3′; Y = 5′-UCCGAUGC-3′
b. X = 5′-AGGCTACG-3′; Y = 5′-CGUAGCCU-3′
c. X = 5′-GCATCGGA-3′; Y = 5′-UCCGAUGC-3′
d. X = 5′-GCATCGGA-3′; Y = 5′-AGGCUACG-3′
e. X = 5′-CGTAGCCT-3′; Y = 5′-CGUAGCCU-3′

13. Which one of the following describes the central dogma of molecular biology?
a. Genetic information flows from DNA through RNA to protein.
b. Genetic information flows from DNA through ribosomes to protein.
c. DNA can be used as a template to make proteins.
d. RNA can be used as a template to make proteins.
e. All of the above.

14. Which one of the following events occurs during the <u>initiation</u> stage of protein synthesis in eukaryotes?
a. Peptide bond formation catalyzed by peptidyl transferase
b. Incorporation of methionine (Met)
c. Movement of the growing polypeptide chain from the P site to the A site

d. Binding of a release factor
e. Separation of the small and large ribosomal subunits

15. Which one of the following statements is incorrect?
    a. The nucleotide sequence in an mRNA molecule is translated in a 5′ to 3′ direction.
    b. The genetic code uses a single codon to specify more than one amino acid.
    c. The genetic code uses triplets of nucleotides to specify amino acids.
    d. With a few exceptions, the genetic code is universal among all organisms.
    e. How the bases in mRNA are grouped into codons (the "reading frame") for a particular protein is determined by the initiator codon.

16. Which one of the following statements about ribosomes is incorrect?
    a. Ribosomes are complexes of ribosomal RNAs (rRNAs) and proteins.
    b. Ribosomes are found in the nucleus in eukaryotic cells and in the cytoplasm in bacterial cells.
    c. Ribosomes are composed of one large and one small subunit.
    d. A single protein-coding sequence in an mRNA can be translated by multiple ribosomes simultaneously.
    e. Protein synthesis occurs on ribosomes.

17. Which one of the following statements about transfer RNA (tRNA) molecules is incorrect?
    a. tRNA molecules act as "bridges" between amino acids and the mRNA template.
    b. tRNA molecules contain bases other than the normal bases, A, G, C, T and U.
    c. tRNA molecules are encoded by genes that are transcribed by RNA polymerase III in eukaryotes.
    d. Cells need more than 64 different tRNA molecules.
    e. tRNA molecules can form unusual (noncomplementary) base pairs with the base in the third position of mRNA codons.

18. Which of the following statements about translation (protein synthesis) is incorrect?
    a. Formation of the initiation complex, binding of a charged tRNA molecule, and peptide bond formation are the three steps that repeat during the elongation phase of protein synthesis.
    b. During elongation, charged tRNA molecules bind to the A site of the ribosome.
    c. Anticodons on tRNA molecules base pair to codons in the mRNA molecule.
    d. During initiation, the charged initiator tRNA molecule base pairs to an AUG codon in the P site.
    e. Termination of protein synthesis does not involve tRNA molecules.

19. The nucleotides CAT were paired with the nucleotides GTA. You could say with some degree of certainty that this pairing <u>most likely</u> did not occur during
    a. DNA replication.
    b. transcription.
    c. translation.
    d. transcription or translation.
    e. any of the above.

20. Which of the following mutant proteins would be <u>least likely</u> to result from a base substitution mutation?

a. A truncated or shortened protein
b. An inactive (nonfunctional) protein with one different amino acid
c. An active (functional) protein with one different amino acid
d. An active protein that contains exactly the same amino acids as the protein made from the unmutated (wild type) gene
e. A protein with a single extra amino acid

21. Use the codon table provided below to determine which one of the following amino acids could not result from a base substitution mutation occurring in the codon AAG.
a. Glu
b. Thr
c. Arg
d. Ile
e. Met

	U	C	A	G	
U	Phe Phe Leu Leu	Ser Ser Ser Ser	Tyr Tyr Stop Stop	Cys Cys Stop Trp	**U** **C** **A** **G**
C	Leu Leu Leu Leu	Pro Pro Pro Pro	His His Gln Gln	Arg Arg Arg Arg	**U** **C** **A** **G**
A	Ile Ile Ile Met	Thr Thr Thr Thr	Asn Asn Lys Lys	Ser Ser Arg Arg	**U** **C** **A** **G**
G	Val Val Val Val	Ala Ala Ala Ala	Asp Asp Glu Glu	Gly Gly Gly Gly	**U** **C** **A** **G**

22. A protein coding sequence contains three codons that specify the amino acid arginine (Arg-1, Arg-2 and Arg-3). In a particular base substitution mutant, Arg-1 is replaced by the Trp codon. In another base substitution mutant, Arg-2 is replaced by the Met codon and in yet another base substitution mutant Arg-3 is replaced by an isoleucine codon. Given this information, use the codon table provided to determine which one of the following statements is incorrect.
a. Arg-1 could be CGG or AGG.
b. Arg-3 = AGA.
c. A base substitution mutation in the Arg-3 codon could give rise to a Stop codon.
d. Arg-2 = AGG.
e. A base substitution mutation in the Arg-1 codon could give rise to a Phe codon.

23. Which of the following statements describes the phenotype of the partial diploid (merodiploid) strain?

$lacI^{+}\ lacP^{+}\ lacO^{C}\ lacZ^{+}\ lacY^{+}\ lacA^{+}/lacI^{-}\ lacP^{+}\ lacO^{+}\ lacZ^{+}\ lacY^{+}\ lacA^{+}$

a. The *lac* gene products would be made continuously or constitutively in this cell.
b. The *lac* gene products would only be made in the presence of lactose.
c. The *lac* gene products would only be made in the presence of glucose.
d. The *lac* gene products would only be made in the presence of lactose and glucose.
e. None of the above statements is an accurate description of the phenotype of the partial diploid strain described above.

24. The <u>negative</u> control system that governs the lactose operon in *E. coli*
a. involves the formation of a cAMP-CRP complex which enhances RNA polymerase binding.
b. involves the binding of a repressor protein to an operator sequence within the *lac* promoter region.
c. operates at the level of translation.
d. necessitates the addition of both lactose and glucose in order for *lac* gene expression to occur.
e. All of the above accurately describe negative regulation of the lactose operon.

25. *E. coli* cells with which one of the following genotypes would be able to produce normal (active) beta-galactosidase in the absence of lactose? [*lacP* = the promoter for the *lac* operon.]
a. $lacI^- \ lacP^- \ lacO^+ \ lacZ^+ \ lacY^+ \ lacA^+$
b. $lacI^- \ lacP^+ \ lacO^+ \ lacZ^+ \ lacY^+ \ lacA^+$
c. $lacI^- \ lacP^+ \ lacO^+ \ lacZ^- \ lacY^+ \ lacA^+$
d. $lacI^+ \ lacP^+ \ lacO^C \ lacZ^- \ lacY^+ \ lacA^+$
e. $lacI^+ \ lacP^+ \ lacO^+ \ lacZ^+ \ lacY^+ \ lacA^+$

26. Which one of the following statements best describes the phenotype that would be exhibited by the following merodiploid (partial diploid) strain in the <u>presence of glucose</u>?

$$lacI^- \ lacP^+ \ lacO^+ \ lacZ^+ \ lacY^+ \ lacA^+/lacI^- \ lacP^+ \ lacO^C \ lacZ^+ \ lacY^+ \ lacA^+$$

a. The *lac* genes would be expressed at a high level (efficiently) in the absence of lactose.
b. The *lac* genes would be expressed at a high level only when lactose is present.
c. The *lac* genes would be expressed at a low level (inefficiently) in the absence of lactose.
d. The *lac* genes would be expressed at a low level only when lactose is present.
e. The *lac* genes would not be expressed.

27. Which one of the following statements about the lactose or *lac* operon is correct?
a. The *lac* operon produces a protein that is needed to transport beta-galactosidase out of the cell.
b. The *lac* operon is an example of a repressible operon because the operon is "turned off" in response to the availability of lactose in the environment.
c. In the presence of lactose, transcription of the *lac* operon is blocked by the *lacI* repressor protein.
d. The *lac* operon is under the control of a weak promoter that is recognized very inefficiently by RNA polymerase.
e. The *lacI* repressor protein can only bind to the operator sequence in the presence of the cAMP-receptor protein (CRP).

28. The product of the *lacI* gene
a. can only bind to an operator sequence that is located on the same DNA molecule.

b. enhances expression of the *lac* genes.
c. binds to glucose.
d. transports lactose into the cell.
e. can bind to operator sequences on distant DNA molecules.

29. An *E. coli* mutant cannot synthesize beta-galactosidase, but produces lactose permease continuously in the absence of lactose. Which one of the following *E. coli* mutants could account for this phenotype?
a. An *E. coli* strain with a mutation in the *lacZ* gene
b. An *E. coli* strain with a mutation in the *lacI* gene
c. An *E. coli* strain with a mutation in the *lac* operator sequence
d. An *E. coli* strain with 2 mutations—one in the *lacZ* gene and the other in the *lacI* gene
e. An *E. coli* strain with a mutation in the *lac* promoter region

30. An *E. coli* mutant produces beta-galactosidase continuously in the absence of lactose (constitutive expression). This mutant strain is converted into a merodiploid by the insertion of a plasmid carrying a wild type copy of the *lac* operon and *lacI* gene. The resulting merodiploid continues to produce beta-galactosidase in the absence of lactose. Based on this information, you could conclude that the mutation is most likely in the
a. *lac* operator sequence.
b. *lacI* gene.
c. *lac* promoter.
d. All of the above
e. a or b

31. Which one of the following statements describes the phenotype that would be exhibited by the following merodiploid (partial diploid) strain in the absence of glucose?

$$lacI^{+}\ lacP^{+}\ lacO^{C}\ lacZ^{-}\ lacY^{+}\ lacA^{+}/lacI^{+}\ lacP^{-}\ lacO^{C}\ lacZ^{+}\ lacY^{+}\ lacA^{+}$$

a. Beta-galactosidase would be produced at low levels in the absence of lactose.
b. Beta-galactosidase would be produced at high levels in the absence of lactose.
c. Beta-galactosidase would not be made.
d. Beta-galactosidase would be produced at high levels only when lactose is present.
e. Beta-galactosidase would be produced at low levels only when lactose is present.

32. Which one of the following mechanisms of gene regulation is present in bacteria but not in eukaryotes?
a. Operons
b. Alternative splicing
c. Cell-specific gene regulatory proteins
d. Regulatory proteins that decrease gene expression
e. Regulatory proteins that enhance gene expression

33. In bacteria, some metabolic pathways are regulated by increasing or decreasing the activity of a key enzyme involved in the pathway. This is an example of
a. transcriptional control.
b. posttranscriptional control.
c. translational control.
d. allosteric regulation.
e. All of the above

34. A repressor protein binds to a DNA sequence and decreases transcription of a gene that is located thousands of nucleotides away. This DNA sequence is most probably
  a. A promoter.
  b. An operator.
  c. An enhancer.
  d. A silencer.
  e. An upstream promoter element.

35. A regulatory protein binds to a DNA sequence and increases transcription of a gene that is located thousands of nucleotides away. This DNA sequence is most probably
  a. A transcription factor.
  b. An operon.
  c. An enhancer.
  d. A silencer.
  e. An upstream promoter element.

36. Gene X is expressed only in muscle cells and gene Y is expressed only in fibroblast cells. You want to express gene X in fibroblast cells. You could probably do this by
  a. moving gene X into fibroblast cells.
  b. replacing the regulatory regions (for example, enhancers and upstream promoter elements) for gene Y with those for gene X in fibroblast cells.
  c. replacing the regulatory regions for gene X with those for gene Y in muscle cells.
  d. isolating the muscle cell-specific regulatory proteins that control the expression of gene X in muscle cells and introducing them into fibroblast cells.
  e. isolating the fibroblast cell-specific regulatory proteins that control the expression of gene Y in fibroblast cells and expressing them in muscle cells.

37. In eukaryotes, all of the following mechanisms of gene regulation play a direct or indirect role in increasing or decreasing the production of RNA by RNA polymerase, except
  a. enzymes that acetylate histones.
  b. enzymes that methylate bases in DNA.
  c. general transcription factors.
  d. alternative splicing.
  e. gene amplification.

38. Which one of the following statements about genetic regulation in eukaryotes is incorrect?
  a. In eukaryotes, genetic regulation can occur between transcription and translation.
  b. In eukaryotes, the expression of cell-specific genes typically involves one or more cell-specific regulatory proteins.
  c. In eukaryotes, the expression of cell-specific genes typically involves one or more cell-specific regulatory sequences.
  d. In eukaryotes, genetic regulation occurs most commonly at the level of transcription.
  e. In eukaryotes, the levels of gene expression can be increased or decreased by sequences that are located thousands of bases away from the genes they affect.

39. Your red blood cells and white blood cells perform different functions because
  a. different genes are expressed in each of these cell types.
  b. red blood cells contain regulatory proteins that are not present in white blood cells.
  c. different mRNAs are produced in each of these cell types.

d. different proteins are produced in each of these cell types.
e. All of the above

40. Which one of the following statements about restriction enzymes is incorrect?
    a. Restriction enzymes cut single-stranded and double-stranded DNA molecules at specific sequences.
    b. A restriction fragment length polymorphism or RFLP often results from the loss of a recognition sequence for a restriction enzyme.
    c. The sequences recognized by most restriction enzymes are palindromes.
    d. The addition of methyl groups to bases in the recognition sequence protects the bacterial genome from its own restriction enzyme.
    e. Restriction enzymes are produced naturally by bacteria to break down foreign DNA that enters the cell.

41. Two type II restriction enzymes, *Dra*I and *Hpa*II are isolated. *Dra*I recognizes the sequence TTTAAA and *Hpa*II recognizes the sequence GGCC. When genomic DNA from a particular bacterium was treated with these two enzymes, it was observed that the bacterial DNA was cut many times by *Dra*I but not at all by *Hpa*II. The <u>most likely</u> explanation for this observation is that
    a. certain environmental conditions may exclude the use of guanines and cytosines in some DNA molecules.
    b. the site for *Dra*I has a greater probability of being found in any one particular DNA molecule than the site for *Hpa*II.
    c. the genomic DNA was isolated from the bacterium that makes *Hpa*II and therefore had been methylated to prevent *Hpa*II from restricting the bacterium's own DNA.
    d. the bacterial DNA specifically inactivated the *Hpa*II enzyme.
    e. None of the above could explain this observation.

42. The following two plasmids were cut with *Eco*RI, the resulting fragments mixed together, and DNA ligase added.

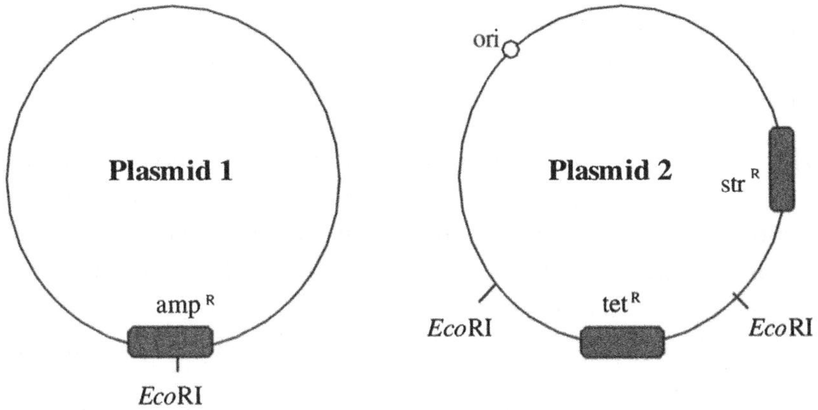

[The superscript R means that a gene confers resistance to that particular antibiotic, where amp = ampicillin, tet = tetracycline, and str = streptomycin. If an antibiotic resistance gene is not shown in a plasmid, assume that the plasmid is sensitive (S) to that particular antibiotic. The lines intercepting the plasmid indicate where the adjoining restriction enzyme cuts the plasmid.] Using the restriction data provided in the figure, determine which one of the following antibiotic resistance phenotypes could not be exhibited by bacteria transformed with the resulting ligation products assuming that each bacterium picks up only one plasmid product.

a. amp^R str^S tet^S
b. amp^S str^R tet^R
c. amp^S str^R tet^S
d. amp^S str^S tet^R
e. amp^R str^R tet^S

43. A genomic library
   a. is a good source of a eukaryotic gene if the purpose of cloning that eukaryotic gene is to produce the eukaryotic protein in a bacterial host cell.
   b. is a collection of recombinant plasmids containing all of the coding sequences of a particular organism.
   c. is a collection of recombinant plasmids containing all of the coding and noncoding sequences of a particular organism.
   d. is a list of all of the genes present in a particular organism.
   e. is made by using the enzyme reverse transcriptase to make DNA copies of the genes that you want to study.

44. In reverse transcription, ____________ is made from ____________.
   a. mRNA; DNA
   b. complementary DNA; mRNA
   c. complementary DNA; DNA
   d. RNA; mRNA
   e. complementary DNA; primary RNA transcripts

45. Which one of the following statements about complementary DNAs (cDNAs) is incorrect?
   a. cDNAs are produced by using the viral enzyme reverse transcriptase.
   b. cDNAs are typically much shorter than the genes that they represent.
   c. cDNAs are directly prepared from primary RNA transcripts.
   d. Extracts from different tissues yield different collections of cDNAs.
   e. cDNAs are prepared if the goal of the cloning is to produce a eukaryotic protein in a bacterial cell.

46. Which one of the following statements about plasmids is incorrect?
   a. Antibiotic-resistance genes on plasmids can be used to isolate bacteria that carry a copy of a plasmid from bacteria that lack a copy of the plasmid.
   b. Plasmids are found naturally in eukaryotes where their role is to degrade any foreign DNA that enters the cell.
   c. Plasmids can be introduced into bacterial host cells by a process known as transformation.
   d. The presence of antibiotic-resistance genes on plasmids has contributed to the emergence of bacteria resistant to multiple antibiotics.
   e. Commercially available plasmids often contain a polylinker that allows for the easy insertion of a piece of DNA into the plasmid.

47. Which one of the following statements about the polymerase chain reaction, or PCR, is incorrect?

a. The PCR primers 5′-AGCCAT-3′ and 5′-TAGTCG-3′ could be used to amplify a DNA fragment flanked by the following sequences:

5′-AGCCAT -------------------------- CGACTA-3′
3′-TCGGTA -------------------------- GCTGAT-5′

b. The polymerase chain reaction requires single-stranded DNA templates, primers, dNTPs (dATP, dTTP, dGTP, dCTP), dideoxy analogs of the 4 dNTPs (ddATP, ddTTP, ddGTP, ddCTP), and a heat-stable DNA polymerase.
c. The polymerase chain reaction can be used to recover sequences from minute amounts of starting material, even from a single cell.
d. In the polymerase chain reaction, the following three steps are repeated over and over: denaturation, primer annealing, and DNA replication (or primer extension).
e. The polymerase chain reaction doubles the amount of the target DNA sequence with each cycle.

48. Pick the pair of primers that you could use to amplify the following double-stranded DNA sequence by PCR:

5′-TAAAGGCTGCAGTTAGAGTGGAAACGCAAGCTCTATGCC-3′
3′-ATTTCCGACGTCAATCTCACCTTTGCGTTCGAGATACGG-5′

a. 5′-TAAAGGCTGCA-3′ and 5′-CCGTATCTCGA-3′
b. 5′-TAAAGGCTGCA-3′ and 5′-GGCATAGAGCT-3′
c. 5′-ATTTCCGACGT-3′ and 5′-CCGTATCTCGA-3′
d. 5′-ATTTCCGACGT-3′ and 5′-GGCATAGAGCT-3′
e. 5′-TAAAGGCTGCA-3′ and 5′-ATTTCCGACGT-3′

49. A restriction map
a. shows the locations of genes in a DNA molecule.
b. shows where restriction enzymes cut a DNA molecule relative to other restriction sites in the DNA.
c. is a listing of all known restriction enzymes and the sequences that those enzymes recognize and cut.
d. relies on gel electrophoresis to determine the sizes of the resulting restriction fragments.
e. b and d

50. The dideoxy DNA method of sequencing DNA
a. relies on termination of DNA synthesis at every position along the DNA template by dideoxy analogues of the four normal deoxyribonucleoside triphosphates.
b. usually requires some knowledge of the nucleotide sequence flanking the 3′ end of the fragment to be sequenced so that a primer can be made.
c. involves four different sequencing mixtures.
d. relies on gel electrophoresis to separate DNA fragments that differ in length by a single nucleotide.
e. All of the above

51. The nucleotide sequence of a 15 bp DNA fragment was determined by the dideoxy DNA sequencing method. The data are shown below.

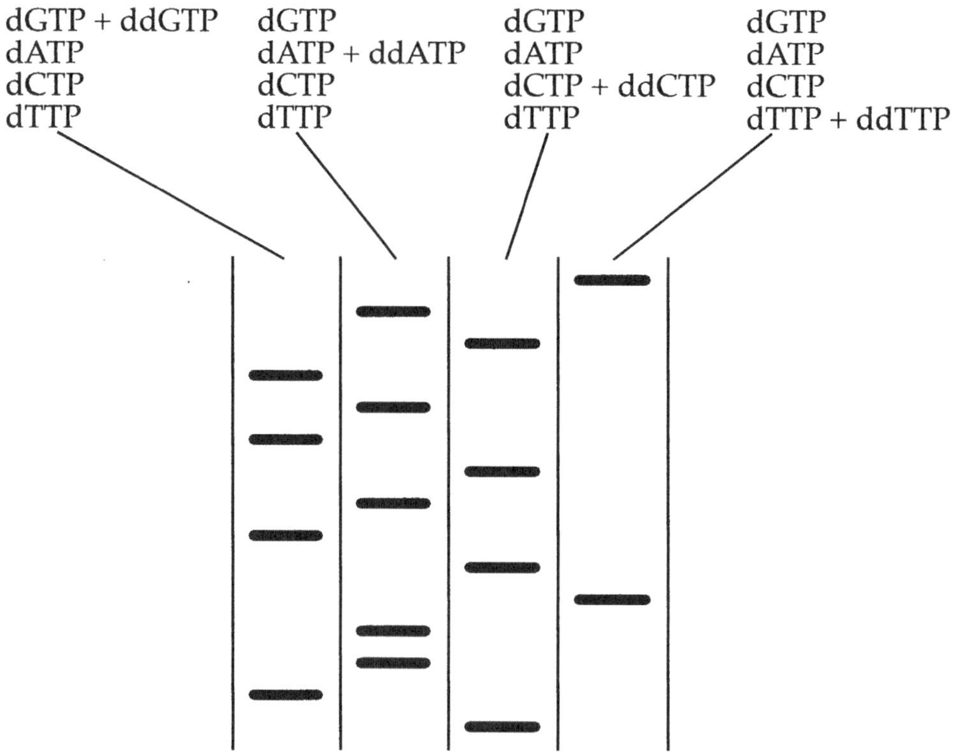

Which one of the following options is correct, where:

X = the 5′ to 3′ sequence of the newly-synthesized DNA strand, and
Y = the 5′ to 3′ sequence of the template DNA—the template DNA is the single-stranded DNA that is added to the sequencing reactions at the start.

a. X = CGAATCGACGAGCAT; Y = GCTTAGCTGCTCGTA
b. X = CGAATCGACGAGCAT; Y = ATGCTCGTCGATTCG
c. X = TACGAGCAGCTAAGC; Y = GCTTAGCTGCTCGTA
d. X = GCTTAGCTGCTCGTA; Y = ATGCTCGTCGATTCG
e. X = CGAATCGACGAGCAT; Y = TACGAGCAGCTAAGC

52. Which of the following statements about recombinant DNA techniques is correct?
a. Blunt-ended fragments isolated from two different DNAs can be joined together even if different restriction enzymes were used to cut each DNA.
b. Most eukaryotic genes have to be converted into complementary DNAs (cDNAs) if the purpose of the cloning is to produce the eukaryotic protein in bacteria.
c. Eukaryotic genes have to be fused to bacterial promoter sequences for expression in bacteria because bacterial RNA polymerase cannot recognize eukaryotic promoter sequences.
d. A DNA fragment can be used to detect complementary DNA sequences using Southern blotting or DNA-DNA hybridization techniques.
e. All of the above statements accurately describe genetic engineering techniques.

53. Match the following terms with the most suitable description:

1. DNA fingerprint
2. Southern blotting
3. Polymerase chain reaction
4. Cloning
5. Recombinant DNA

a. The act of making many identical copies of a DNA fragment inside a host cell.
b. Base pairing of a radioactively-labeled probe DNA to complementary sequences immobilized on a solid support.
c. A DNA molecule consisting of combined DNAs that originated from two different sources.
d. Unique restriction digest DNA profile.
e. The amplification of a specific DNA sequence in the presence of primers.

a. 1 = d, 2 = b, 3 = a, 4 = e, 5 = c
b. 1 = d, 2 = b, 3 = e, 4 = a, 5 = c
c. 1 = c, 2 = b, 3 = a, 4 = e, 5 = d
d. 1 = d, 2 = b, 3 = e, 4 = c, 5 = a
e. 1 = c, 2 = b, 3 = e, 4 = a, 5 = d

Final Exam: Answers

1. **T** 2. **F** 3. **T** 4. **F** 5. **T** 6. **F** 7. **T** 8. **c** 9. **e** 10. **b** 11. **d** 12. **c**

13. **a** 14. **b** 15. **b** 16. **b** 17. **d** 18. **a** 19. **d** 20. **e** 21. **d** 22. **e** 23. **a**

24. **b** 25. **b** 26. **c** 27. **d** 28. **e** 29. **d** 30. **a** 31. **c** 32. **a** 33. **d** 34. **d**

35. **c** 36. **d** 37. **d** 38. **c** 39. **e** 40. **a** 41. **c** 42. **e** 43. **c** 44. **b** 45. **c**

46. **b** 47. **b** 48. **b** 49. **e** 50. **e** 51. **b** 52. **e** 53. **b**

INDEX

Printed and bound by CPI Group (UK) Ltd, Croydon, CR0 4YY
09/06/2025
14685997-0005